Cold Fusion: The Scientific Fiasco of the Century

John R. Huizenga is the Tracy H. Harris Professor
of Chemistry and Physics at the University of Rochester.
He was Co-Chairman of the United States Department of Energy –
Energy Research Advisory Board Cold Fusion Panel.

Cold Fusion: The Scientific Fiasco of the Century

JOHN R. HUIZENGA

UNIVERSITY OF ROCHESTER PRESS

First published 1992

University of Rochester Press
200 Administration Building, University of Rochester
Rochester, New York 14627, USA
and at PO Box 9, Woodbridge, Suffolk IP12 3DF, UK

ISBN 1 878822 07 1

Library of Congress Cataloging-in-Publication Data
Huizenga, John R. (John Robert), 1921–
    Cold fusion : the scientific fiasco of the century / John R.
Huizenga.
        p.   cm.
    Includes bibliographical references and index.
    ISBN 1–878822–07–1   (acid-free)
    1. Cold Fusion. I. Title.
    QC791.775.C64H85   1992
    539.7'64–dc20   91–47072

British Library Cataloguing-in-Publication Data
Huizenga, John R.
    Cold fusion: the scientific fiasco of the century.
    I. Title
    539.7
    ISBN 1–878822–07–1

This publication is printed on acid-free paper

Printed in the United States of America

CONTENTS

To
DOLLY
LINDA, JANN
ROBERT, JOEL

PREFACE

In the Spring of 1989, two electrochemists promised the world an energy utopia – clean, cheap and abundant energy without harmful side effects to the environment. B. Stanley Pons of the University of Utah and Martin Fleischmann of Southampton University announced that they had successfully created a sustained nuclear fusion reaction at room temperature in a small jar on a laboratory tabletop. They had duplicated the process powering the sun. Their reported accomplishment had eluded fusion scientists for several decades, in spite of the fact that these scientists were experimenting with extremely high temperatures and large machines, and spending billions of dollars in fusion research. Fleischmann and Pons indeed made heady promises, which if fulfilled, are the stuff of Nobel prizes.

It was the scientific story of the century broadcast around the globe, and made network television newscasts on March 23. For example, the respected MacNeil/Lehrer News Hour devoted extended coverage to cold fusion and included interviews with both Fleischmann and Pons. Scientists around the country examined the videotapes of the MacNeil/Lehrer News Hour and other broadcasts to search for tidbits of information. However, the experimental details necessary for scientific evaluation were missing from the press release and early reports, leading many to be both upset by Fleischmann and Pons' "publication-by-press-conference" and skeptical of their claims.

The dream of cold fusion had a natural appeal because it was fueled by a very special set of circumstances. The environmental and political issues in the United States had focused our attention on a lack of a clear national energy policy. The almost coincident Exxon Valdez oil disaster was uppermost in the public's mind. Fears over gasoline shortages, price rises and lines at service stations were prevalent. The national consciousness concerning energy was rising dramatically. Furthermore, the burning of fossil fuels was known to produce large amounts of carbon dioxide in the air adding to the Earth's greenhouse gases and their predicted effect on global warming. In addition, the burning of our rather abundant high-sulfur coal was producing acid rain with its devastating consequences.

These highly destructive characteristics of fossil fuel on our environment led many to rethink the use of nuclear fission energy. However, the nuclear waste problem and the remembrance of the nuclear catastrophe of Chernobyl dampened the prospect of moving to large-scale usage of nuclear fission power in the near future. Others held nuclear fusion to be the energy source of the future. Billions of dollars had been spent already on controlled fusion research and development. Although considerable progress has been made in both magnetic and inertial fusion, no commercial power plants are projected to be operating, if at all, until well into the middle of the next century. It is against this rather dismal portrayal of the world's future energy supply that Fleischmann and Pons came onto the scene and stunned the world with their claim to have solved the controlled fusion problem. No wonder the promise of cheap, clean and abundant energy captured everyone's imagination.

In the early euphoric days of cold fusion, disbelief in the new energy dream was unpopular and viewed almost as an unpatriotic act. The *Wall Street Journal* compared Fleischmann and Pons with Ernest Rutherford, one of the giants of twentieth-century physics. However, with over four decades of experience in nuclear science, I was skeptical, as were most of my immediate colleagues, of Fleischmann and Pons' spectacular claims. Enhancing the probability of a nuclear reaction by 50 orders of magnitude (10^{50}) via the chemical environment of a metallic lattice, contradicted the very foundation of nuclear science.

Even so, many of us moved quickly to participate in the verification process. Surprises do occasionally occur in science. In scientific research it is always important to be on the lookout for an unexpected or surprising result. Our research group at the University of Rochester had state-of-the-art neutron detectors and associated electronics. This instrumentation was immediately mobilized by several of my colleagues to search for evidence of room temperature nuclear fusion. Experimentation is the final authority in science and experimental groups around the world immediately attempted to verify test-tube fusion on a bench top as viewed on the evening news. Electronic networks were saturated with open questions and unconfirmed rumors about cold fusion.

Early in April, 1989 I received a telephone call from John Schoettler, Chairman of the Energy Research Advisory Board (ERAB), asking me to serve as chairman of an ERAB panel on cold fusion. ERAB, an advisory committee to the Secretary of the Department of Energy, often formed study panels on issues of interest to the Secretary. I had served on ERAB since 1984 and felt some obligation to accept, but asked for a short time to consider the implications of such an appointment. The next week I was in Dallas and attended the first large public session on cold fusion organized by the American Chemical Society. This most unusual meeting, dubbed the "Woodstock of Chemistry", demonstrated the sharp division between scientists on the reality of cold fusion. All indications were that there would be no quick resolution of Fleischmann and Pons' extraordinary claim. A few days later, following discussions with colleagues at the spring meeting of the National Academy of Sciences, I agreed to co-chair the DOE/ERAB panel.

The panel members were appointed immediately by ERAB and began their work in late April.

For the next six months, I was completely immersed in the study of cold fusion. As the co-chair of the DOE/ERAB panel, I was in a unique position to participate directly in the day to day exchange of claims and counter claims and to experience the excitement and mystery associated with the cold fusion saga. Confirmations, retractions, new positive claims and null results were the order of the day and all had to be distilled and evaluated. It was the responsibility of our panel to gather up-to-date information from every group in the United States, as well as from many foreign groups, researching cold fusion. The amount of material was voluminous. Teams of panel members also made visits to selected laboratories working on cold fusion. Our panel completed its interim report in July, 1989 and final report in November, 1989. On speaking to a number of different groups and organizations about our panel's conclusions and recommendations, I learned firsthand that many people had a deep curiosity about the whole cold fusion episode and wanted to learn more about it. This motivated me to write this book on "cold fusion."

It is important for the reader at the outset to understand that the term "cold fusion" (or "room temperature fusion") has frequently been used for two very different phenomena. Firstly, the University of Utah claimed that Fleischmann and Pons had "successfully created a sustained nuclear fusion reaction at room temperature" producing four watts of power for each watt of input power. Secondly, Professor Steven E. Jones and his physicist colleagues at Brigham Young University (BYU) reported that they had produced very low levels of neutrons from fusion at room temperature. These two claims differed by thirteen orders of magnitude (ten thousand billion). Even so, both claims have been labelled "cold fusion" and are often erroneously interpreted as the same phenomenon.

This interchangeable use of "cold fusion" for these two very different claims has added considerable confusion to the cold fusion saga. The claim of watts of excess heat from nuclear fusion of deuterium at room temperature is completely inconsistent with reports placing very low limits on the intensities of fusion products. Jones' claim is not an independent corroboration of Fleischmann and Pons' claim. In this volume I focus on the more exotic claim of high fusion rates, first reported by Fleischmann and Pons. It is this claim that has excited the interests of everyone and has promised to solve our energy needs for all time. I do discuss also, however, the much more modest claim of very low levels of fusion products from deuterium fusion at room temperature. The reader, therefore, must always be aware of the fundamental difference in the two claims. From a scientific point of view, the latter claim, if true, is extremely interesting in its own right, but this phenomenon has no practical potential as an energy source.

The first six chapters deal with events through the second month of the cold fusion saga. Once the promise of room temperature fusion had been announced by press conference, without first passing through the normal processes of scientific scrutiny, the verification process entered the public arena. These chapters describe the early reports on cold fusion mostly communicated at several scientific meetings in the presence of the media. The clash between science and the politics of science is an integral part of the cold fusion episode. The University of Utah lobbied in Washington for major funding before the science had even been confirmed.

The second group of chapters (VII to X) describe and evaluate some of the pertinent scientific data. The University of Utah in its original press release called the experiment "extremely simple." In opposition to this, definitive calorimetric experiments turned out to be very difficult. This was especially true for open cell calorimeters. The positive reports were plagued by experimental uncertainties, inadequate controls and improper assessment of errors. One group of very strong proponents of excess heat summarized its results at the First Annual Conference on Cold Fusion by stating "There are enough calibration runs which show too much heat and D_2O runs which show little or no heat [so] that the whole process could be noise."

If fusion of deuterium is occurring there must be tell-tale fusion products. The detection sensitivity for fusion products is orders of magnitude larger than that for excess heat. Therefore, searches for neutrons, tritium, helium, etc. are the key experiments to validate cold fusion. These are described and analyzed in some detail. Proponents agree with skeptics on at least one aspect of the cold fusion saga. Namely, that there is an extremely large disparity between the claimed amounts of excess heat and fusion products even as reported by proponents. True believers in cold fusion have been frustrated by this incongruous result because it undermines the very foundation of the promise of a new "clean virtually inexhaustible source of energy." In order to believe simultaneously in the claimed large amount of excess heat on the one hand and in its nuclear origin on the other hand, believers resorted to pseudoscience. Conventional nuclear physics was declared invalid in metallic lattices by fiat. This opened the door for a succession of miracles such as excess heat without fusion products and tritium without neutrons. Research teams obtaining negative results were often characterized as part of the "eastern establishment" and dismissed with the barb, "negative results can be obtained without skill and experience." The proponents claim of "new physics in solids" has added intrigue and hype to the cold fusion saga, but unfortunately, it has in the final analysis led to confusion, scandal and deception. Fleischmann and Pons' underlying reason for investigating room temperature fusion was flawed from its very inception. They mistakenly asserted that the pressures attained during electrolysis were sufficient to drive deuterium nuclei close enough to fuse. The National Cold Fusion Institute has closed, careers have been damaged and many tens of millions of dollars have been squandered in time and resources. Still no verification.

How did cold fusion germinate and what fueled the whole episode? Is cold fusion pathological science? What are the hazards of going public with a far-reaching promise without sufficient experimental evidence? These subjects are explored in the final three chapters. A majority of scientists were unable to replicate Fleischmann and Pons' claim. Still, over a hundred groups reported excess heat and/or some fragmentary evidence for trace amounts of fusion products. On the basis of the sheer number of positive claims, it is tempting to conclude, as many believers have, that there must be some truth to cold fusion. Numbers of unproven claims alone, however, are not definitive in science. Hundreds of papers were published in support of both N rays and polywater, both classic examples of pathological science, which was defined by Irving Langmuir, Nobel laureate in chemistry, as "the science of things that aren't so." The paranoia of the advocates of cold fusion is illustrated by their charge that a highly vocal small group with hot fusion interests are sabotaging the future development of cold fusion. This is a case of self-deception, a characteristic of pathological science.

The University of Utah's handling of cold fusion is a striking illustration of what happens when administrators use potential royalties to force premature publication and when universities lobby for large federal funds before the science is confirmed. The chimera of cold fusion with excess heat is a striking illustration of what happens when research is done in isolation by scientists who are outside their field of expertise, when scientists circumvent the normal peer review process, when scientists require too many miracles to account for their results, when data are published by others through private communication rather than by the researchers responsible, when scientists distort the normal scientific procedures to protect patent rights, and when scientists use the press as a conduit to disseminate information about a claimed discovery in an unrealistic and overly optimistic tone.

The cold fusion fiasco illustrates once again that the scientific process works by exposing and correcting its own errors.

John R. Huizenga
Rochester, New York

ACKNOWLEDGMENTS

I wish to thank publicly members of the Energy Research Advisory Board Cold Fusion Panel (see Appendix II for names) for their dedicated work during the six months that we studied intensively the voluminous literature on cold fusion coming from laboratories around the world. This group of twenty-two scientists from diverse backgrounds and with different fields of interest converged in their thinking on cold fusion to produce reports that were unanimously agreed on by all members. The following panel members deserve to be singled out for their special contributions to our final report: Allen J. Bard, Jacob Bigeleisen, Howard K. Birnbaum, T. Kenneth Fowler, Richard L. Garwin and John P. Schiffer. Without the support and encouragement provided by all the panel members during our study, I would not have later made the decision to write this book. However, I should emphasize that the opinions and conclusions expressed here are my own and are not necessarily those held presently by other panel members. My thanks also go to panel staff members Thomas G. Finn, David Goodwin and William Woodward who were very helpful to us in all phases of our work.

The panel also enjoyed the whole hearted backing from our parent body, the Energy Research Advisory Board. They were the authoritative body that unanimously approved our panel's final report for submission to the Secretary of Energy, Admiral James D. Watkins.

I am indebted to Tim Fitzpatrick for supplying photocopies of his stories in the *Salt Lake Tribune*, Bob Welk for placing cold fusion news clips from the *Wall Street Journal* in my mailbox, Bruce V. Lewenstein for sending me a copy of the University of Utah press release and numerous friends for calling particular articles to my attention. I am especially grateful to a large number of my scientific colleagues who supplied me with preprints of their articles and reports on cold fusion, as well as assorted bits of information by BITNET and FAX. The articles entitled "Cold fusion News" written and electronically distributed by Douglas R.O. Morrison are classic pieces on cold fusion.

Several people have read early versions of this manuscript and provided helpful comments and suggested corrections. These include my daughter, Jann, my wife, Dolly, Nathan S. Lewis, Jack A. Kampmeier, Douglas R.O. Morrison and W. Udo Schröder. I greatly appreciate the help that each has given me. I also thank Dolly for sitting through the Hearing before the Committee on Science, Space and Technology on April 26, 1989 in Washington and taking prolific notes.

ACKNOWLEDGMENTS

I particularly wish to thank Debbie Shannon-Mryglod for her special help in typing and preparing the book for publication. Thanks are also due to editors Robert Easton and Pam Cope for their helpful suggestions.

ABBREVIATIONS

ACS	American Chemical Society
AIP	American Institute of Physics
APS	American Physical Society
BARC	Bhabbha Atomic Research Center
BITNET	electronic mail
BNL	Brookhaven National Laboratory
BYU	Brigham Young University
Bull. Am. Phys. Soc.	*Bulletin of the American Physical Society*
Caltech	California Institute of Technology
CBS	Columbia Broadcasting System
C&EN	*Chemical and Engineering News*
CERN	Conseil European pour la Recherche Nucleaire (large particle-physics laboratory in Geneva)
Coll. Jour. USSR	*Colloid Journal of the USSR*
COSEPUP	Committee on Science, Engineering and Public Policy
DOE	United States Department of Energy
EPRI	Electric Power Research Institute
ERAB	Energy Research Advisory Board
Europhys. Lett.	*Europhysics Letters*
FAX	facsimile machine
GANIL	Grand Accélérateur National d'Ions Lourds
IBM	International Business Machines
J. Electroanal. Chem.	*Journal of Electroanalytical Chemistry*
J. Electrochem.	*Journal of Electrochemistry*
J. Phys. Chem.	*Journal of Physical Chemistry*
J. Radioanal. Nucl. Chem. Letters	*Journal of Radioanalytical Nuclear Chemistry Letters*
keV	kilo-electron volts
mA	milli-amperes
MIT	Massachusetts Institute of Technology
MeV	Million electron volts
NAS	National Academy of Science
NASA	National Aeronautics and Space Administration
NBC	National Broadcasting Company
NCFI	National Cold Fusion Institute

NSF	National Science Foundation
Phys. Lett.	*Physics Letters*
Phys. Rev.	*Physical Review*
Phys. Rev. Lett.	*Physical Review Letters*
PNL	Pacific Northwest Laboratory
Proc. Natl. Acad. Sci.	*Proceedings of the National Academy of Sciences*
Proc. Roy. Soc.	*Proceedings of the Royal Society (London)*
SIMS	secondary ion mass spectroscopy
SDI	Strategic Defense Iniative
Soviet Tech. Phys. Lett.	*Soviet Technical Physics Letters*
SRI	Stanford Research Institute
UCLA	University of California Los Angeles
WKB	Wentzel-Kramers-Brillouin approximation
ZETA	Zero Energy Thermonuclear Assembly
Z. Naturforsch.	*Zeitschrift für Naturforschung*
Z. Phys.	*Zeitschrift für Physik*

I

Press Conference

On March 23, 1989, two electrochemists, Dr. B. Stanley Pons, Chairman of the Chemistry Department at the University of Utah, and Dr. Martin Fleischmann, a Research Professor at Southampton University, reported a major breakthrough in nuclear fusion research during a startling press conference in Salt Lake City. They claimed to have produced nuclear fusion in a test tube at room temperature in a high-school-type apparatus. If true, it was the type of discovery for which Nobel prizes are awarded! The accompanying press release (see Appendix I) was general in nature and contained virtually no technical information. Most of the major network and cable news programs carried positive stories about the Fleischmann-Pons phenomenon which has been dubbed "cold fusion." These media reports raised the hopes and expectations of many people. For example, Dan Rather led off the CBS Evening News that night with a fusion report, exclaiming, "What may be a tremendous scientific advance" (*Time*, April 8, 1989, p. 74). Journalists came to Salt Lake City from all over the United States and Europe to cover the story. The *Wall Street Journal* made cold fusion the top story in its world-wide news column on March 24. The page-one story flashed the headline, "Taming the H-Bombs?". This highly positive and detailed article in America's most prestigious financial journal suggested to the world that fusion of deuterium had actually been accomplished at room temperature. The *Wall Street Journal* article in explaining cold fusion stated that in the palladium metal lattice "the deuterium nuclei are brought close enough together to overcome their mutual repulsion and fuse." For informed readers this was one of the most extraordinary and bizarre claims made in the thirty-five-year-long effort to produce a controlled, sustained hydrogen-fusion reaction. Although most other news organizations emphasized the importance of the University of Utah discovery as a solution to the world's energy problems, some major news organizations did express skepticism and gave the story only a minor coverage. The *New York Times*, for example, relegated the story on March 24 to page A16 and wisely included reactions by physicists and fusion scientists who were skeptical of the University of Utah claims. *USA Today* for some reason ignored the story completely in its March 24 edition.

A very interesting story developed several hours prior to the University of Utah press conference. *The Financial Times* of London broke the cold-fusion story on the morning of the U.S. press conference, surprising even the University of Utah press office. This occurred because March 24, 1989 was Good Friday, a bank holiday in England. Due to this holiday, *The Financial Times* would not be published on March 24, and if the paper were forced to wait until the March 23 press conference in Utah it wouldn't be able to run the cold-fusion story until Monday, March 27. This delay was apparently unacceptable to Fleischmann. According to one published report [*Science* **244**, 422 (1989)] Fleischmann contacted an old friend, Richard Cookson, to ask him about the best way to get good coverage in Britain. Cookson, a former colleague of Fleischmann in the Chemistry Department at the University of Southampton, put Fleischmann in touch with his son Clive Cookson, who writes for *The Financial Times*. As a result of this contact, Fleischmann, with Pons' approval, provided *The Financial Times* with the information about their discovery for publication a day early. Hence, *The Financial Times* got the "scientific story of the year" one day before anyone else (*Science* **244**, 422). The *Wall Street Journal* also had a story on the morning of March 23 entitled "Development in Atom Fusion to be Unveiled." However, its story provided only background information for its March 24 story which followed the University of Utah's press conference.

It has been known for many years that nuclear fusion occurs under exotic conditions of high temperature and pressure, such as are prevalent in the interior of the sun. Therefore, the harnessing of fusion energy for commercial use has been an elusive dream for many decades. The Fleischmann-Pons claim of cold nuclear fusion gave the world the promise of the century, namely, the promise of a virtually limitless supply of a cheap, safe and environmentally clean nuclear energy. If true, this would be an extraordinary accomplishment. Up to 1989, scientists in several countries, including the United States, USSR, Japan and a consortium of European countries had spent billions of dollars in fusion research, working on experiments with the extremely high temperatures believed necessary for inducing fusion. It is no wonder then that Fleischmann and Pons' claim shocked the scientific world and led the public to believe that the world's energy problems might be solved for all time.

The creators of the ensuing fusion frenzy first met in Southampton, England. Pons graduated from Wake Forest University in 1965, spent two years in graduate school at the University of Michigan, and left to join his family in the textile business. After eight years in business, Pons yearned to return to science and was accepted into the Electrochemistry Department at the University of Southampton to pursue his doctorate, which he obtained in 1978. It was there that he met Fleischmann, at that time a professor of electrochemistry. The two became friends and later, after

Figure 1. Professors Martin Fleischmann (right) and B. Stanley Pons in their University of Utah Chemistry Laboratory. Here they claimed to have successfully created a sustained nuclear fusion reaction at room temperature in a simple apparatus on a tabletop. Pictured are four of their small electrolysis cells in a constant-temperature water bath. At the Dallas American Chemical Society meeting (see Chapter III), Pons dubbed this bench-top experiment "the U–1 Utah Tokamak". (Courtesy of the University of Utah.)

3

Fleischmann took early retirement from Southampton University, they worked on many joint projects at the University of Utah and published their findings together.

According to some sources (e.g., *Time*, May 8, 1989, p. 76), the idea of cold nuclear fusion was born some five years prior to the press conference during one of Fleischmann's visits to Utah. During a hike up Millcreek Canyon in Utah's Wasatch Mountains, the pair became convinced that they were onto something big. According to Pons "We'd stayed up all night, trying to figure the experiment out, and it just wasn't coming together. So we went for a walk in the canyon to clear our minds and pretty soon things started coming together" (*Salt Lake Tribune*, April 30). This article stated that "after returning to the house, Pons and Fleischmann opened a bottle of bourbon and worked through the night once again. By the next morning the bourbon bottle was nearly empty and the experiment was ready to be played out in the laboratory." According to Pons "Within eight months, we'd obtained significant results", "but a 'meltdown' accident forced us to scale back our experiments. It was incredibly frustrating, like starting from scratch." There is controversy about the exact time when this meltdown actually occurred. *Time* magazine (May 8, 1989) reported the 'meltdown' occurred one night in 1985. Pons' evaluation of the incident was "we had much more energy than could be attributed to a chemical reaction." For the purpose of assigning priority and patent rights, it was in Fleischmann and Pons' interest to establish the earliest possible date for their first "successful" experiment.

The University of Utah claims were particularly astounding given the simplicity of their equipment, just a small test-tube-like cell with a pair of electrodes connected to a battery and immersed in a jar of heavy water, D_2O. Heavy water is a form of ordinary water H_2O in which the abundant light isotope of hydrogen with atomic mass number one (^1H) is replaced by the rare heavy isotope of hydrogen with atomic mass number two (^2H). In this manuscript I will refer to the heavy isotope as deuterium and use the symbol D rather than ^2H. Deuterium was discovered in 1932 by the re-nowned chemist Harold Urey. Since deuterium is an isotope of hydrogen it has very similar chemical properties with the ordinary isotope of hydrogen. There is sufficient deuterium (the amount of D in the universe is approxi-mately 0.015% of the total hydrogen) in seawater to power the globe for essentially an infinite time, provided some way could be found to overcome the mutual repulsion of the positively charged deuterium nuclei to allow nuclear fusion to occur. The electrodes in the Fleischmann-Pons cell were made of palladium, the negatively-charged electrode (cathode), and a metal such as platinum, the positively-charged electrode (anode). An electrolyte was added to the heavy water in the cell to make the solution electrolyti-cally conductive. Although the chemical formula of the electrolyte was not made public immediately, word spread within a few days that it was lithium deuteroxide, LiOD. During the electrolysis of the deuterated solution, the electric current decomposes the heavy water into deuterium gas (D_2) and oxygen gas (O_2), and drives large amounts of deuterium (D) into the palladium cathode. The idea that palladium, or some other metal like titanium, might catalyze fusion stemmed from the special ability of these metals to absorb large quantities of hydrogen (or deuterium). Fleischmann and Pons conjectured that during the electrolysis process the deuterium atoms, within the confines of the palladium metal lattice, would be in close enough proximity to enhance markedly their probability of fusion. In fact, they reasoned that the probability would be enhanced so much as to allow measurable yields of nuclear fusion at room temperatures. It was the elementary nature of Fleischmann and Pons' experiment, along with the media's reports of the potential significance of their discovery, that captured everyone's imagination and raised Fleischmann and Pons to celebrity status overnight.

Statements in the original press release (see Appendix I) and in follow-up news stories emphasized especially the simplicity of the reported experiments. For example, Fleischmann and Pons had "achieved nuclear fusion in a test cell simple enough to be built in a small chemistry laboratory" (New York Times, March 24); "Our indications are that the discovery will be relatively easy to make into a usable technology for generating heat and power" (Science, March 31); "They have produced a sustained fusion reaction for 100 hours" (Salt Lake Tribune, March 24); "Evidence that fusion was taking place was the fact that in addition to heat they detected the production of neutrons, tritium and helium – the expected by-products of fusion reactions" (Wall Street Journal, March 24). James Brophy, the vice-president for research at the University of Utah, reinforced the view of the experiment's simplicity by his statement that "the experiment is easy to carry out once you know how. Fleischmann and Pons have reproduced it dozens of times" (Austin American-Statesman, March 28).

Some early support for the Fleischmann and Pons claim came from Edward Teller, Director Emeritus of the Lawrence Livermore National Laboratory. Teller, who has often been called the "father of the hydrogen bomb", said that what he had heard of the experiment sounded promising. Although Teller said he was skeptical and of the "opinion that cold fusion could never happen" before a reporter read him the text of Thursday's (March 23) news release, he stated after the press conference that "I am extremely happy now because I see a very good chance that I was completely wrong" (Salt Lake Tribune, March 24). Very soon after the press conference, Teller phoned Pons for additional technical information and was one of a select few persons to have his call completed. Teller was favored with a preprint of the paper that Fleischmann and Pons had submitted to the Journal of Electroanalytical Chemistry (see Chapter II).

As illustrated above, most of the early stories were very positive and reflected excitement over a perceived major discovery of the century. At the core of the March 23 press conference was the striking claim that in a simple electrochemical cell at room temperature, four watts of heat were produced from an input of one watt of electricity. The assertion, that the excess heat[1] was due to a nuclear reaction process, caused a number of scientists, almost immediately after the University of Utah press release, to point out fundamental inconsistencies in the reported accomplishments and to ask other hard questions about the discrepancies between the reported production of heat and the amounts of fusion products.

The process of fusion of two deuterium atoms was first studied by Ernest Rutherford and his colleagues at the Cavendish Laboratory in Cambridge (*Nature*, **133**, p. 413, 1934; *Proc. Roy. Soc.* **A148**, p. 623, 1934). During the ensuing half century, nuclear fusion of deuterium has been studied intensively and it is now a relatively well understood process. If fusion between deuterium atoms were indeed occurring at room temperature, there is essentially no doubt what the products would be. They would be the same products that are observed for reactions between two low-energy deuterium nuclei, where fusion is known to proceed in three ways:

$$D + D \rightarrow [^4He]^* \rightarrow \begin{cases} ^3He\ (0.82\ MeV) + n\ (2.45\ MeV) & \textbf{(1a)} \\ T\ (1.01\ MeV) + p\ (3.02\ MeV) & \textbf{(1b)} \\ ^4He\ (0.08\ MeV) + gamma\ ray\ (23.77\ MeV) & \textbf{(1c)} \end{cases}$$

The mechanism of such reactions in going from the reactants (in this case the two deuterium nuclei on the left) to the products (right side) was first elucidated by Niels Bohr, the great Danish theoretical physicist, in the 1930s. In his seminal contribution, he showed that fusion reactions occur in two separate stages. In the first stage, the reactants fuse into an excited intermediate nucleus, ($[^4He]^*$), identified as the Bohr "compound nucleus."

---

[1] More precisely, excess power. However, in this manuscript, I have used heat and power interchangeably. In electrolysis experiments, the input power (watts) is defined as the product of the current (amperes) and the potential difference (volts), $P(watts) = I(amperes) \times V(volts)$. Energy (joules) is defined as power (watts) multiplied by time (seconds), $E(joules) = P(watts) \times t(seconds)$. Most experiments are measuring and reporting excess power. Excess power at some time or other during the electrolysis is nothing out of the ordinary and does not necessarily mean that the charged palladium-deuterium system has produced excess energy. A power burst could represent a time-dependent release of energy previously stored such as by a battery which is suddenly connected by a switch to a starter motor. Excess energy requires that the output power integrated over time is larger than the input power integrated over time. It is, hence, necessary to determine whether or not the cell is producing excess energy, not power.

The asterisk is used to denote a nucleus with internal excitation energy. In the second stage, the intermediate nucleus deexcites by decaying into products. The most important property of the Bohr compound nucleus is that its lifetime is sufficiently long to separate the two stages of the reaction. Hence, the decay of the compound nucleus is entirely independent of how it was formed, accounting for factors such as different angular momentum distributions.

The reactions (1a) and (1b) have been studied over a range of deuteron kinetic energies down to a few kiloelectron volts (keV) and the cross sections (production rates) for these two reactions have been found experimentally to be nearly equal (to within ten percent). Hence, the fusion of deuterium produces approximately equal yields of 2.45 million-electron-volts (MeV) neutrons (with an accompanying ^{3}He atom) and 3.02–MeV protons (with an accompanying tritium atom). This near-equality of the neutron and proton branches (production rates) is expected also on the basis of theoretical arguments. The cross section (production rate) for reaction (1c) is several orders of magnitude lower than reactions (1a) and (1b). This well-established experimental result is consistent with the Bohr model which predicts that the compound nucleus will decay predominantly by particle emission [reactions (1a) and (1b)], as opposed to radiative capture [reaction (1c)], whenever it is energetically possible.

Based on present experimental evidence, the branching ratios for the three reactions shown by Equation (1) appear to be essentially constant at low energies. There is no reason to think that these branching ratios would be measurably altered for cold fusion. Therefore, if the Fleischmann-Pons claim of watts of excess heat were due to fusion between deuterium nuclei, large amounts of easily detectable fusion products[2] would have to be present. In fact, one watt of power from nuclear reactions (1) must have associated with it approximately 10^{12} (one million million) neutrons per second, neutrons being the fusion product most easily detected by direct counting. Early critics recognized that on the basis of the reported heat, Fleischmann and Pons would have to had been exposed to massive, lethal doses of nuclear radiation. They were, however, in good health and had obviously not been exposed to such lethal radiation dosages. The fusion product yields as reported by Fleischmann and Pons later turned out to be some one hundred million times less than expected from the power reported at the March 23 press conference, avoiding such a blatant inconsistency. However, an enormous discrepancy remained between the claims of heat production and the failure to observe commensurate levels of fusion products, which are by far the most sensitive signatures of fusion. This striking incon-

---

[2] Even if the branching ratios for cold fusion were very different, large amounts of fusion products would still have to be present.

sistency led many, early on, to be doubtful of the claim that nuclear fusion had been induced at room temperature and was the source of the reported heat. If the heat were due to a nuclear reaction it was essential, based on conservation laws, to have detected the commensurate amounts of fusion products. The lack of this balance between the heat and fusion products made Fleischmann and Pons' claim of high levels of nuclear fusion unbelievable to most scientists.

A second reason for early skepticism was the major inconsistency between the Fleischmann-Pons results and those obtained by a group at a neighboring university. Fifty miles from the University of Utah, a group of physicists, headed by Professor Steven E. Jones, at Brigham Young University (BYU), also claimed cold nuclear fusion, based on an extremely small yield of measured neutrons. If translated into energy, their result would be over 10^{12} (a million million) times smaller than the watt level of heat claimed by Fleischmann and Pons. Jones made the interesting comparison between the BYU and the University of Utah results in terms of dollars by saying that if the BYU results were assumed equivalent to a dollar, the University of Utah results would exceed the national debt. Jones was known for his earlier research on muon-catalyzed fusion, a type of low-energy fusion well understood theoretically. His group also had some experience with the detection of very low intensities of neutrons, hence, the results of Jones et al. tended to be more believable by nuclear scientists. Even so, the Jones et al. neutron results were barely above background,[3] leading many to be skeptical also of their claimed positive identification of fusion. Although the BYU results, if true, were of great scientific interest, the corresponding energy output from their electrolytic cells would have been extremely small and of no practical interest as far as commercial energy production is concerned. Jones et al. neither measured nor claimed any excess heat from their experiments.

The third reason for skepticism is that cold fusion should not be possible according to current nuclear theory, which is supported by a large body of experimental data. Deuterons are positively charged and repel each other by the electromagnetic force, which has a long range. Hence, fusion reactions can occur only if during a nuclear collision the Coulomb barrier is surmounted or, at low energies, at least penetrated so that the nuclei approach each other within about 10^{-5} nanometers. A nanometer is one-billionth of a meter (10^{-9} meter). At this close distance the short-range, but powerful, nuclear force can cause fusion. The challenge is to bring the deuterons close enough for fusion, a distance that is some ten thousand

[3]   The background is due to neutrons or neutron-like events arising from sources *other* than D+D fusion [see reaction (1a)], for example, from cosmic rays and natural radioactivities.

Figure 2. Brigham Young University physicists are pictured with their neutron detection equipment. From left to right are Professors Steven E. Jones, J. Bart Czirr, Gary L. Jensen, Daniel L. Decker and E. Paul Palmer. Their electrolytic cells and neutron spectrometer are shielded with many twenty-five-dollar boxes of pennies. This group made no heat measurements, however, first reported very small intensities of neutrons resulting from cold fusion experiments. In explaining the enormous difference between the magnitude of their claim and that of the University of Utah electrochemists, Jones illustrated with dollars. Assuming the BYU fusion product intensity was equivalent to a dollar, the amount of excess heat claimed by the University of Utah would exceed the national debt. (Courtesy of Professor S.E. Jones.)

times smaller than the typical separations of atoms in ordinary matter. For example, the deuterium-deuterium (D–D) distances in a palladium metal lattice range from 0.28 to 0.17 nanometers, depending on whether the deuterium occupies only octahedral sites or a combination of octahedral and tetrahedral sites. Molecular-dynamical-simulation calculations at high deuterium to palladium concentrations indicate that even under dynamic conditions, the D–D distances are not shorter than the molecular D–D distance of 0.074 nanometers. Hence, the likelihood that D+D fusion occurs at room temperature in a palladium lattice is extremely small.

The penetration of the Coulomb barrier at low temperatures takes place through a well-understood quantum mechanical phenomenon called tunneling, which allows fusion to occur in collisions far less violent than might be required otherwise. The cross section for fusion varies markedly with energy, due to the fact that the probability for penetration of the repulsive Coulomb barrier changes rapidly with deuteron kinetic energy. The theoretical rate for D+D cold fusion in diatomic deuterium molecules, as calculated by Steven E. Koonin and Michael Nauenberg, is $10^{-63.5}$ fusion events per second per pair of deuterium atoms. Koonin and Nauenberg are professors of physics at the California Institute of Technology (Caltech) and the University of California, Santa Cruz, respectively. To put the above number in perspective, it corresponds to one fusion event per year for a solar mass of deuterium! No mechanism is known by which this rate could be enhanced by nearly 50 orders of magnitude (ten times itself fifty times) to agree with the University of Utah claim of watts of excess heat.[4] Hence, it would be extremely surprising if claims were true according to which the penetration of the barrier by quantum-mechanical tunneling had been observed in the laboratory at room temperature. Such a claim must be viewed with extreme skepticism.

These three major difficulties caused a number of scientists to be very skeptical of the most unusual claims made by Fleischmann and Pons during their March 23 press conference. Their concerns, however, were initially only rarely quoted by the news media. Instead, most commentaries expressed enormous optimism about our future energy supply. The dream of having abundant energy with little pollution was so captivating that reasoned arguments were put aside. When confronted with the fantastically important claim to have harnessed nuclear fusion – the power of the sun – in a test tube of D_2O at room temperature, it is understandable that scientists all over the world rushed to their laboratories in order to repeat this "unbelievable" accomplishment. Hence, the above-mentioned difficulties were temporarily put aside and the scientific race of the century was in-

---

4 If such an enhancement did miraculously occur, the fusion products would have to be present in an amount commensurate with the claimed heat.

itiated. After all, it was argued that scientific claims of this magnitude can, in the final analysis, only be settled quickly by additional experiments.

In the thermonuclear fusion process that occurs in stars and in laboratory "hot fusion" experiments, high temperatures (more than tens of millions of degrees) provide the violent collisions required to produce fusion. That fusion does occur under such conditions was firmly established with the successful development of thermonuclear weapons. As early as 1929, several years before fusion reactions were first observed in the laboratory, Robert Atkinson and Fritz Houtermans proposed the now well-accepted explanation that fusion is the source of energy for the sun.

The secrecy surrounding Fleischmann and Pons' experiments on campus prior to the March 23 press conference caused major problems. The secrecy prevented open debate and discussion. This can be best illustrated by the fact that the University of Utah scientists and administrators did not even consult members of their own physics department before announcing cold fusion to the world. To have made an announcement, of what appeared at that time to be a major scientific breakthrough in nuclear fusion, without seeking expert opinion from the physicists on campus was a serious intellectual and administrative mistake. When physicists and chemists around the United States learned through the news media about the claims of Fleischmann and Pons, they contacted their physicist and chemist colleagues at the University of Utah. It was with total disbelief that the scientific community outside Utah learned that the University of Utah physicists had been completely unaware of both the impending announcement and the cold fusion research on their own campus! No doubt the rivalry between the University of Utah and Brigham Young University, known to be near publication of its own cold fusion work, closed the debate between the two scientific groups and pressured the University of Utah officials into a premature press conference. Even so, it is hard to understand how university officials could go public on such a major claim, involving nuclear physics at its deepest level, without consulting and involving the physicists on campus. This neglect is even more striking because James Brophy, vice president for research and chief spokesman for the administration, is a physicist by training, and knew of the discrepancy between the expected nuclear products and the findings of Fleischmann and Pons.

Calling a press conference is not the conventional way in science to announce a new scientific discovery. Such a procedure circumvents the normal peer review process. This is one of the most important steps in the scientific publication process whereby one's new findings are subjected to a close examination by experts in your field. This process serves to eliminate the more obvious mistakes that might have been made, and, in addition, peers often see the necessity for additional evidence to support one's conclusions prior to approving publication. There have been precedents, however, when new results are important enough to justify an early press conference. When choosing this mode of presentation, the researchers have to be very convinced that their results are correct and be able to back up their results by all the conceivable experimental checks. Although Fleischmann and Pons were perhaps genuinely convinced that cold fusion had in fact occurred in their experiments, it will be shown later that they did not carry out a number of even the more elementary tests and cross-checks.

II

Prior Events

The March 23, 1989 press conference did not represent the first time that nuclear fusion had been claimed to occur at room temperature. The earliest such claim was in the late 1920s in experimental studies whose conclusions were later retracted. Two German scientists, F. Paneth and K. Peters, published a paper in *Die Naturwissenschaften* [**14** 956 (1926)] in which they reported the transformation of hydrogen into helium by spontaneous nuclear catalysis at room temperature occurring whenever hydrogen is absorbed by finely ground palladium metal. The theory of the day could not explain their results, but there was good reason to attempt at least the transformation of hydrogen into helium to produce a safer lifting gas for airships. The results of Paneth and Peters caused great excitement at the time. In this respect, the 1926 episode exhibits striking parallels to the current furor over cold fusion.

The spectroscopic measurement of the minute amounts of helium (as small as one-thousand millionth of a cubic centimeter) by Paneth and Peters was by far the most challenging part of the experiment. They performed numerous control experiments to check for possible errors in the interpretation of their results. After some time, Paneth and Peters showed that the liberation of helium from glass is dependent on the presence of hydrogen. Glass tubes, which gave off no detectable helium when heated in vacuum or in an oxygen atmosphere, yielded up absorbed helium in the quantities observed when heated in an atmosphere of hydrogen [*Nature* **119** 706 (1927)]. Hence, they were in a position to give an explanation of the occurrence of the observed very small quantities of helium in their experiments as coming from absorbed helium in the glass, without having to conclude the synthesis of helium from hydrogen. These authors then acknowledged that the helium they measured was due to background from the air, and published a retraction from their earlier claim.

Little was understood about thermonuclear fusion in 1926, although Paneth and Peters mentioned in their paper the hypothesis that helium is produced from hydrogen in stars. In 1927, Swedish scientist J. Tandberg extended the Paneth and Peters experiments by using the electrolysis of water to get hydrogen into a palladium electrode. He applied for a Swedish patent for "a method to produce helium and useful reaction energy." This invention was an electrolytic cell, using ordinary water. Tandberg's device, based on the pioneering work of Paneth and Peters, was claimed to have a significant increase in efficiency. The patent, however, was never granted. After the discovery of deuterium in 1932, Tandberg continued his electrolysis experiments with heavy water (D_2O) in place of light water (H_2O) [ordinary water, e.g. sea water, contains 99.985% H_2O and only a minute 0.015% D_2O]. Hence, the experimental setup of Tandberg had many similarities to that used by Fleischmann and Pons some sixty years later. Although Tandberg had high expectations of producing helium and energy from his electrolysis experiments, he failed on both counts. His work was carried out in the Stockholm laboratory of Electrolux, a manufacturer of household appliances, where he spent many years as a scientific director and manager.

In March, 1951 Argentina President, Juan Perón, declared that one month earlier "thermonuclear experiments were carried out under conditions of control on a technical scale" in a pilot plant at the secret Huemul Island Laboratory on Lake Nahuel Huapi near the Pategonian town of San Carlos de Bariloche (*Physics Today*, January, 1953, page 21). This was the largest laboratory of experimental science in Argentina and was at that time under the direction of the German-educated fusion expert, Dr. Ronald Richter, and four of his German and Austrian associates. Richter, himself an Austrian refugee, was a nuclear physicist in Germany under Hitler. He later moved to Argentina and convinced Perón that he knew how to build a controlled fusion reactor. The location of Richter's laboratory is known for its Swiss-like scenery and its mountains, lakes and forests. In the July, 1952 issue of *Physics Today* (page 5), Dr. G.L. Brownell reported on his visit to Richter's laboratory that was devoted entirely to thermonuclear research. Due to security, Brownell and the other United States visitors were limited to discussions with Richter outside the experimental areas. Richter told the visitors that their work was aimed solely towards the production of energy for industrial purposes, and expressed his optimism that concrete results would be forthcoming in the near future. Perón had told the people of Argentina that a decision had been made not to invest in the enormous capital expense necessary to produce nuclear fission. Instead Perón made the conscious decision to take the risk and pursue thermonuclear fusion. With this background, Perón proudly reported that controlled thermonuclear fusion was attained by his fusion guru Richter! In backing Richter, Perón put a lot of personal and national prestige on the line. Although causing considerable initial excitement in the press, the story was treated by most scientists with intense skepticism. One skeptic expressed the general scientific mood of the time by stating "I know what the material is that the Argentines are using . . . It's baloney." On December 4, 1952 Edward R.

Murrow reported that the whole project of which Perón boasted was completely discredited, the staff of 300 scientists at the laboratory sent home, and Richter was arrested. After Perón was dismissed from the government, the new government found that 70 million dollars had been wasted on futile efforts to generate energy.

The next major chapter in fusion research was written in 1956 when Luis W. Alvarez of the University of California at Berkeley, an outstanding experimental physicist and a pioneer of bubble-chamber research, made a remarkable discovery. While he and his colleagues were examining bubble-chamber data, they observed some strange particle tracks. These tracks were due to muons that had stopped inside a chamber filled with liquid hydrogen and deuterium, and represented the first experimental observation of muon catalyzed fusion. In his Nobel-prize acceptance speech in 1968 Alvarez stated "We had a short but exhilarating experience when we thought we had solved all of the fuel problems of mankind for the rest of time." A negative muon is an electron-like particle but 207 times more massive than an electron. It is the muon's large mass that enables it to catalyze fusion reactions.

The possibility that negative muons could catalyze nuclear fusion in hydrogen isotopes had first been suggested and studied by F. C. Frank of the University of Bristol in England in the late 1940s. These ideas were further developed shortly afterward by Andrei D. Sakharov and other Soviet physicists. Alvarez and his colleagues were unaware of this theoretical work in 1956 when making their experimental discovery. Negative muons form very strong bonds between the nuclei of hydrogen atoms. Because the muon is 207 times more massive than the electron, it can follow an orbit about 207 times smaller than the electron's. The muon-bonded hydrogen nuclei then fuse, ejecting the muon, which can go on to catalyze other fusion reactions. Since the muon has a decay lifetime of only about two microseconds and may also be captured by other heavier parasitic nuclei in still shorter times, it is important to catalyze as many fusion reactions as possible before the muon disappears. Jones and his collaborators in the early 1980s, working with mixtures of deuterium and tritium, measured some 150 fusions per muon by varying the temperature.

Is muon-catalyzed fusion a practical source of energy? Presently, the answer is NO, because the cost of producing muons is too great. Although muon-catalyzed fusion is often called cold fusion, it is a well understood process, and will not be included in our discussion of cold fusion at room temperature. In any further reference, this type of fusion will be termed "muon-catalyzed" fusion.

In 1958, two years after the discovery of muon-catalyzed fusion, Britain's Sir John Cockcroft stunned the world with an announcement of having harnessed nuclear fusion. He stated that he was "90% certain" that his Zero

15

Energy Thermonuclear Assembly (ZETA) had produced a controlled fusion reaction between two deuterons (D+D). The **mighty ZETA**, wrote the *London Daily Mail*, will allow the world to have "Limitless fuel for millions of years." Unfortunately, Cockcroft's optimistic statement was incorrect. Once again the dream of cheap, abundant energy had fizzled out.

The experiments of Fleischmann and Pons followed a long history of unsuccessful attempts to achieve controlled fusion of hydrogen isotopes. The University of Utah Press reported that Fleischmann and Pons had intensely pursued their research on cold fusion for five years prior to going public on March 23, 1989. During this early period, while their research on cold fusion was performed in isolation from the scientific world, it has been reported, Fleischmann and Pons spent $100,000 of their own money (later refunded by an anonymous donor). When examining the record, it is evident that during the five-year period prior to 1989, Pons and Fleischmann had major commitments to activities unrelated to cold fusion. For example, Pons was chair of the University of Utah's chemistry department for part of this period and was the director of a major research program in electrochemistry that led to a large number of publications in areas completely outside nuclear fusion. Fleischmann was also a participant in a sizeable fraction of this research. Therefore, the media reports claiming that the two scientists had intensely pursued cold fusion research for five years appear unfounded. More realistically, events indicate that the University of Utah program in cold fusion was first seriously pursued when graduate student Marvin Hawkins joined Pons' research group during the Fall semester of 1988. An unconfirmed BITNET message circulated that Hawkins had claimed the 'meltdown' of the palladium cube occurred in October, 1988. Assigning the 'meltdown' to the Fall of 1988 rather than sometime in 1985 as claimed by the press seems more reasonable. If such a catastrophic event had actually occurred in 1985 and had been interpreted at the time as due to a nuclear reaction, it is difficult to understand why Pons and Fleischmann devoted so much of their effort during this crucial interval of time to their life-long research program in electrochemistry. Isn't it logical that they would have dropped everything and immediately devoted all their future effort to this fantastic discovery?

The intense competition that was to develop between the fusion scientists at the University of Utah and Brigham Young University had its origin in August of 1988. Fleischmann and Pons had prepared a research proposal and submitted it at the end of August to the Advanced Energy Projects Division of the United States Department of Energy (DOE). Jones was selected by the DOE as one of the five referees to evaluate Fleischmann and Pons' proposal. After working on the proposal during part of September, 1988, Jones submitted his report near the end of September. The choice of Jones as a reviewer was quite natural since Jones was a recognized researcher in muon-catalyzed fusion and had initiated work in an area called piezonuclear fusion ('piezo' meaning 'pressure'). Furthermore, he was already funded by the Department of Energy through the Advanced Energy Projects Division. Jones had some reservations about the theoretical material in the proposal. However, he said "I never disapproved of the proposal at any point." Jones in his report to the Department of Energy recommended funding and wrote that the two groups had developed complementary techniques. BYU had in operation a state-of-the-art neutron spectrometer and the University of Utah had expertise in making calorimetric measurements. Jones called Dr. Ryszard Gajewski, project director of the Advanced Energy Projects Division, and suggested that he inform Pons about the BYU research program and, in particular, about their neutron spectrometer. Following Gajewski's contact with Pons, Pons called Jones and requested information on his neutron spectrometer which was supplied. From the outside it seemed that a collaboration between the two groups would have been a natural development. Such a collaboration, however, did not develop for a variety of reasons. Pons claimed he was not interested because he did not believe the BYU group had much to offer. "We never needed Jones' spectrometer, we never wanted his spectrometer." This was a serious mistake in judgment, as will be evident later in the discussion of Fleischmann and Pons' neutron measurements.

Jones had worked on muon-catalyzed fusion since 1981 and continued this work along with the initiation of piezonuclear fusion research after he joined the BYU physics department in 1985. Jones' entry into the piezonuclear fusion field represented an extension of his muon work. His research group was influenced, Jones said, by the work of a Russian scientist named B. A. Mamyrim, who in 1978 reported anomalous concentrations of helium-3 (^3He) in various metals. As early as September, 1986 the BYU group claimed to have very weak evidence of neutrons from cold fusion experiments utilizing a very crude neutron counter. Worried, however, that the neutrons might well be spurious, the BYU scientists embarked on a two-year project to build a neutron spectrometer capable of both counting neutrons and measuring their energies. Late in 1988, the new BYU neutron spectrometer was ready to take data. Some of the experiments with this detector showed a slight excess of neutrons above background in the 2.5–MeV energy region of the spectrum. By early 1989, Jones and his collaborators had sufficient confidence in their data to consider writing a paper.

Following their earlier communications, the University of Utah and BYU groups continued their personal contact, and on February 23, 1989, Fleischmann and Pons came to BYU to talk with the Jones group. During this visit, Jones showed his neutron data to Fleischmann and Pons and indicated that the BYU group felt they had enough information to merit publication of their results. According to statements by Jones, Pons and

Fleischmann indicated that they would prefer to continue to do their research for at least another eighteen months before going public and requested that the Jones group delay reporting their results. This, however, was not possible for Jones. He had already submitted an abstract for an invited paper[5] on his work to be presented at the American Physical Society meeting on May 4 in Baltimore, and he simply couldn't back out of this commitment. Jones did, however, partially acquiesce to Pons and Fleischmann's request by cancelling his planned colloquium in the physics department of BYU. Another interesting discussion at this meeting concerned the possibility of the two groups submitting their research findings at the same time. However, no final agreement was reached at this time on joint publication of their respective research (*Science* **244**, page 28).

Another meeting of representatives of the two neighboring Utah universities occurred on Monday, March 6. Now the scientists Jones, Pons and Fleischmann, joined by the presidents of the two universities, met at BYU. At this meeting University of Utah President Chase Peterson expressed his optimism about the enormous future benefits of cold fusion to the state of Utah and its citizens. This view was at odds with that held at BYU where their results did not suggest any practical energy applications. It was agreed, however, that the two research groups would submit their papers jointly to *Nature* on March 24. Members of each group would meet at the Federal Express Office in Salt Lake City airport to mail their respective manuscripts. Reportedly, Jones and Pons disagreed on the terms that were placed on the agreement, in particular on questions about publicity prior to the time the papers were to be submitted. Jones said "Our understanding was that we would not go public before the submission of the papers", while Pons' version was that "There was never any agreement not to publicize" (*Science* **244**, page 28).

Based on the previous meetings and discussions between members of the University of Utah and BYU and the agreement to publish simultaneously, it is not easy to understand the University of Utah's decision to go public following a rather unusual chain of events. However, I surmise that Fleischmann and Pons were unduly impressed by Jones' neutron data taken with the new BYU spectrometer. This data gave Fleischmann and Pons confidence that neutrons were actually being emitted from their own electrochemical cells. At the time of their February, 1989 meeting with Jones,

---

[5] The deadline for receipt of abstracts was February 3, 1989. Jones' invitation to speak at the APS meeting was based undoubtedly on his research on muon- catalyzed fusion. The content of his abstract was devoted largely to muon- catalyzed fusion, however, it included the following sentence: "We have also accumulated considerable evidence for a new form of cold nuclear fusion which occurs when hydrogen isotopes are loaded into various materials, notably crystalline solids (without muons)".

Fleischmann and Pons were relying on a health physics monitor to detect neutrons! Fleischmann had hoped to have help from Harwell, a major national laboratory in Britain, to confirm that they were actually producing neutrons but, unfortunately, this did not occur for a number of reasons. In a final desperate move to confirm the presence of neutrons, Pons asked R. Hoffman, a health physicist at the University of Utah, in early March, 1989 to monitor his cells with a large sodium iodide detector. The results from these experiments were not only misinterpreted by Fleischmann and Pons but also have raised serious questions about their handling of these data (see Chapters VI and VIII). The worry by the University of Utah adminstrators and scientists that the BYU neutron data would preempt their heat data, and as a result negatively impact on their patent rights and associated billions of dollars in royalties, no doubt played a major role in their decision to jump the gun and submit a paper on their own results.

On March 11, just five days after the agreement on joint publication between the two universities, Fleischmann and Pons submitted a paper to the *Journal of Electroanalytical Chemistry*, unbeknownst to Jones and others at BYU. The Utah paper was received on March 13 in the University of California (Davis) office of W.R. Fawcett, who served as a U.S. Editor of the above Swiss (Lausanne) Journal printed in the Netherlands. Within a week Fawcett had the reviewers' comments back in his office and he and Pons quickly agreed on revisions. The revised manuscript was received in the editorial office ready for the printing process on March 22, a day before the March 23 press conference. The editors apparently believed the manuscript merited this special handling, even though there seemed to be no way to explain or understand the Fleischmann-Pons results. The article appeared in print approximately one month later in the *Journal of Electroanalytical Chemistry*, **261**, 301–308 (1989). Comments on this article will appear in the next Chapter, some of which will address questions related to the advisability of the special treatment this manuscript received in the peer review process.

On March 21, the University of Utah made the critical decision to go public with a press conference on March 23, without informing anyone at BYU. The experimental results of both groups on cold fusion were on the threshold of becoming widely known. One surmises that the University of Utah's decision was based on their concern about establishing priorities, especially as they pertained to Jones' research at BYU. Not knowing the content of the forthcoming BYU paper, Fleischmann and Pons decided to take their case to the popular press. As expected, the University of Utah played down its rivalry with BYU, and emphasized that the principal reason for the press conference was to correct "rumors, leaks, questions and false information" that were circulating. One administrative official went so far as to point to the article in the March 23 *Financial Times* as an example of how the press was closing in (recall that the Times article, initiated by Fleischmann, appeared prior to the Utah press conference).

On March 22, Jones learned that the University of Utah was holding a press conference the next day to announce its discovery. The BYU group was shocked and disappointed and considered this a violation of the agreement worked out on March 6. BYU administrators called the president and the vice-president for research at the University of Utah to protest the announced press conference, however, it occurred anyway. The BYU administrators and scientists were especially offended by the press conference. James Brophy, the vice-president for research at the University of Utah, was asked by a reporter whether he knew of similar work being done at other universities. Brophy answered "We're not aware of any such experiments going on"! This statement was indeed surprising since the BYU scientists had been in direct contact with Fleischmann, Pons and Utah University Administrators, and, in addition, the two groups had agreed previously to publish at the same time. What particularly saddened the soft-spoken Jones was the veiled accusations from the University of Utah that he (Jones) stole ideas from them. "I think, frankly, that there is an effort to discredit us" says Jones (Business Week, May 8, page 103). With the publication agreement broken, Jones and his colleagues decided not to send their manuscript by Federal Express from the Salt Lake City airport on March 24 as originally planned, since they had been tipped off that the press would be present. Instead the BYU group proceeded to fax their manuscript off to Nature so that it would have a March 24 submission date. It is interesting to note that someone from the University of Utah did show up at the airport on March 24 and waited for a delegate from BYU to arrive. This was confirmatory evidence that there had been an agreement for parallel submission of manuscripts.

Both the timing of the University of Utah press conference and the decision by Fleischmann and Pons to publish their existing cold fusion research were dictated largely by factors and circumstances outside usual scientific publication norms. For this reason many scientists criticized the way Fleischmann and Pons publicized their results on cold fusion prior to doing sufficient checks and control experiments. According to reports, Pons and Fleischmann would have preferred to continue their research for at least another eighteen months before going public. For example, Fleischmann himself is quoted as saying that he was not happy with the way things were done: "It's not my normal style of operation." In many ways the quality of Fleischmann and Pons' research results, which were widely circulated, were inconsistent with five long years of intense effort by two competent researchers. Was the University of Utah being carried away with its vision of large financial gains from cold fusion? Many of the University of Utah faculty members had not received a raise in four years, and their salaries were some twenty percent lower than at comparable universities. As a result, a number of faculty had left Utah taking positions elsewhere. If the Utah patents held up and the dream of cold fusion materialized, the University, the Chemistry Department, and Fleischmann and Pons would stand to become very wealthy. It was agreed that each party would receive one-third of the proceedings of the discovery (P. Scarlet, *Salt Lake Tribune*, March 25). To them, cold fusion meant billion dollar patents, institutional and faculty prestige, major new research centers and economic development in the state of Utah.

In concluding this chapter, it is important to emphasize again the great disparity between the claims of the scientists at the University of Utah and Brigham Young University. The University of Utah's claim of watts of excess heat requires commensurate particle intensities that are some 13 orders of magnitude (ten trillion or 10,000,000,000,000 or 10^{13}) greater than the neutron intensity reported by the BYU group. The imbalance between the heat and fusion products reported by Fleischmann and Pons has been a major problem with their claim from the beginning. It is unfortunate that the BYU claim of very small anomalous neutron yields has become somehow associated with the University of Utah claim of large amounts of excess heat and commercial power production. Both claims have been sometimes characterized as cold fusion, but one must obviously be very careful to distinguish between the two very different claims.

21

III

Confirmations, Retractions and Confusion

The political activity following the University of Utah press conference escalated into high gear. Governor N. Bangerter vowed on March 24 to convene a special session of the Utah legislature in order to request five million dollars to ensure that Utah "reaps the rewards of research at the University of Utah that may have unlocked the secret of nuclear fusion." The Governor said "this accomplishment is a result of our previous investment in research and development which has been a high priority during present and past Administrations." Members of the Board of Regents of the University of Utah also expressed pride and excitement in their statements to the press. W. Eugene Hansen, regent chairman, said "when this is proven scientifically sound, it will be one of the most significant events of the century. It was a landmark event." Regent Charles W. Bullen added "this is one of the greatest scientific breakthroughs in the history of mankind." Regent Ian Cumming had more practical advice in pointing out that Fleischmann and Pons' research had "political and economic implications" that higher education and the state ought to pursue. He urged that the university move ahead quickly on the engineering research necessary to make practical use of the local discovery because businesses and governments throughout the world would soon exploit the University of Utah research. Mr. Cumming predicted that "The state of Utah is on the threshold of reaping the benefits of its modest investment in scientific research at the University of Utah." Dr. Chase Peterson, President of the University of Utah, made contacts with prominent Utah industries to generate funding for the necessary engineering research and development. He said "It would be depressing to see Mitsubishi Inc. take off on this in three years because we couldn't move fast enough to capitalize on it" (P. Scarlet, *Salt Lake Tribune*, March 25).

The state of euphoria in Utah following the press conference was exemplified by a proposal made by Utah State Senator Eldon Money. As an alternative to the Governor's plan to request five million dollars from the state legislature, Senator Money proposed that nuclear fusion research in the state be funded with volunteer contributions from taxpayers. In return, the contributors would receive tax breaks when the research paid off.

Senator Money reasoned that "Because of the far-reaching effects and potential of this discovery for creating great economic wealth, it is very important to spread the benefit to as many individuals as possible and not to confine participation to just a few businesses or investors or even to the state of Utah itself" (T. Fitzpatrick, *Salt Lake Tribune*, March 28).

On March 30, James C. Fletcher was reportedly to have been chosen to direct the University of Utah's new National Cold Fusion Institute, to be funded initially by the five million dollars requested from the state legislature (*Salt Lake Tribune*, March 30). Dr. Fletcher, whose ancestors were pioneer Utah settlers, had previously served as president of the University of Utah from 1964 to 1971 and later as the director of the National Aeronautics and Space Administration (NASA) for a number of years. However, this announcement by the University of Utah proved to be premature, as no such appointment has actually materialized. Later, it was announced that Fletcher would serve as advisor on the fusion program to the university's president and to the fusion center, but would continue to make his home in Washington, D.C.

During the immediate period following the press conference, information about the Fleischmann and Pons experiments came principally from the media. Statements made at the University of Utah press conference and in their press release (see Appendix I) emphasized the simplicity and technological importance of the discovery, namely, "A simple experiment results in sustained nuclear fusion . . . the discovery will be relatively easy to make into a usable technology for generating heat and power . . . [T]his generation of heat continues over long periods and is so large that it can only be attributed to a nuclear process." The message of Fleischmann and Pons in late March was very clear: They had produced nuclear fusion in a simple apparatus at room temperature! Once this news broke, *the* scientific race of the century was on.

The announcement made at the March 23 press conference at the University of Utah sent chemists, physicists and other scientists world-wide scurrying to their laboratories to try to reproduce the Utah phenomenon. Scientists entered the race to confirm or shoot down the Utah claims for various reasons. Some did it out of curiosity while many others were dazzled by the prospect of securing a source of virtually limitless clean energy and were covetous of the financial rewards and prestige that could be theirs. The 'science by press conference' atmosphere was fueled by the tantalizing promise of an easy solution to the world's energy problems by an extremely simple technology. Quite uncharacteristic of scientific research all over the world, very early details about the Utah experiments were sketchy at best, and scientists had to work mainly from accounts in the popular press.

Since details of the Fleischmann-Pons experiment were still sketchy at this time, information and rumors were being wildly circulated around the world by electronic mail, such as BITNET, and facsimile (FAX) machines. Unauthorized copies of Fleischmann and Pons' manuscript submitted to the *Journal of Electroanalytical Chemistry* were photocopied and faxed so many times that the only remaining clearly legible words were "Confidential – Do Not Copy." On April 10, the paper by Fleischmann and Pons appeared [*Journal of Electroanalytical Chemistry*, **261**, 301–308 (1989)], just twenty-eight days after submission.

This eight-page paper was followed some weeks later by two pages of errata [**263**, 187–188 (1989)]. Hence, the length of the errata was some 25% of the length of the original paper. When the paper was finally available for examination by an anxious scientific community, most readers were shocked by blatant errors, curious lack of important experimental detail and other obvious deficiencies and inconsistencies. David Bailey, a physicist at the University of Toronto, said the paper was "unbelievably sloppy." He was quoted as saying, "If you got a paper like that from an undergraduate, you would give it an F." He went on to say that the paper "was totally unacceptable" and that "they don't sound like they know what they're talking about." Moshe Gai, a physicist at Yale, added that the paper also made clear that Pons and Fleischmann used "inadequate equipment" and that the nuclear physics part of the experiment "was not done in a very careful way." It is apparent that the speedy publication process circumvented an adequate peer review of the manuscript that would probably have rejected it. Included in the errata was the addition of a third author, M. Hawkins! Forgetting to include one of only three authors of a paper must be a first. Also included in the errata was a complete replotting of the gamma ray spectrum published in the original manuscript with new data points giving a different spectral shape and a change in the ordinate intensity scale of a factor of twenty-five! This spectrum has been the focus of much criticism, to be discussed later.

The Fleischmann, Pons and Hawkins paper stated the following conclusions.

(a) The excess enthalpy (heat) generation is dependent on the applied current density and is proportional to the volume of the electrodes; i.e. we are dealing with a phenomenon in the *bulk* of the palladium electrodes.

(b) Enthalpy generation can exceed ten watts per cubic centimeter of the palladium electrode; this is maintained for experiment times in excess of 120 hours, during which typically heat in excess of four megajoules per cubic centimeter of electrode volume was liberated.

(c) Neutrons are indeed generated in the electrodes . . . [T]he neutron flux . . . is of the order of 4×10^4 per second.

(d) Fig. 1B demonstrates that the species accumulated is indeed tritium . . . and . . . takes place to the extent of $1–2 \times 10^4$ atoms per second which is consistent with the measurements of the neutron flux.

In their original paper, Fleischmann, Pons and Hawkins state further that their data on enthalpy generation would require rates for the reactions (1a) and (1b) (see reactions on p. 6) involving between 10^{11} and 10^{14} atoms per second. They go on to conclude that "it is evident that reactions (1a) and (1b) are only a small part of the overall reaction scheme and that other nuclear processes must be involved." It is this very large discrepancy between the reported production of heat and nuclear products that has caused considerable skepticism in the scientific community. In this article no mention is made of what "other nuclear process" is producing the watts of reported power. Recall that the neutron and tritium production reported by Fleischmann, Pons and Hawkins can account for only approximately one-hundred millionth (10^{-8}) of the reported power. It is one of the great mysteries of the cold fusion episode that the authors proceeded with little or no hesitation to publish their results, claiming that the "bulk of the energy release be due to an hitherto unknown nuclear process or processes." There was not the slightest reservation expressed by the authors that this might be a risky assumption. Their claim did not acknowledge the extensive literature on nuclear reactions acquired and the basic principles established over the last half century. It is perhaps even more mysterious how a paper with such an undocumented and preposterous claim could pass the peer review and editorial processes.

The excess enthalpy (heat) claimed had been measured with a method known as calorimetry. To date, several types of calorimeters have been employed in these and similar studies. Fleischmann, Pons and Hawkins used open isothermal calorimeters from which all gases produced in the electrolytic process are assumed to escape. This particular technique, based on the concept of a leaky thermos, is referred to as heat-leak calorimetry. Some of the process heat is contained in the cell, while the remainder leaks out, at a known rate, into a constant-temperature bath in which the cell is immersed. Sometimes the cell is separated from the constant-temperature bath by a vacuum jacket, in order to control the rate of heat transfer from the cell to the bath. If the cell is a well insulated flask, the heat leak will be very small, and the contents of the electrolytic cell will come to a steady-state temperature that is much higher than that of the constant-temperature bath. If the cell is not well insulated, it will equilibrate rather quickly and be only slightly hotter than the surrounding bath. It is this difference in temperature of the electrolysis cell, T_{cell}, and the bath temperature, T_{bath}, that is monitored. The cell temperature is always higher than that of the constant temperature bath. The relative heat flow from the cell to the bath is determined from the above temperature difference, T_{cell} minus T_{bath}.

In the above type of calorimetry, the cell temperature, T_{cell}, varies and increases and decreases as the power output of the cell increases and decreases. In order to know the heat flow rate on an absolute basis, it is

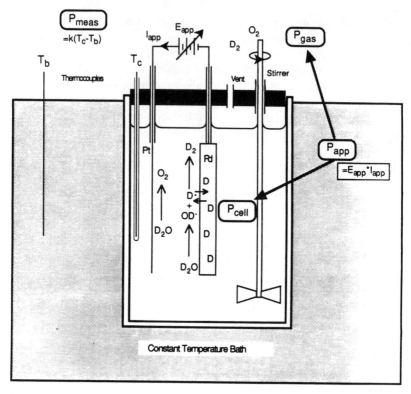

Figure 3. Schematic drawing of isothermal calorimeter used in the California Institute of Technology work [*Science* **246** 793 (1989)]. The cathode is palladium (Pd), which absorbs the deuterium, and the anode is platinum metal (Pt). The power applied through the electrolysis circuit $P_{app} = E_{app} I_{app}$, where E_{app} is the applied voltage and I_{app} is the applied current, produces heating power delivered to the cell, P_{cell}, and to the power contained in the evolved gases, P_{gas} ($P_{app} = P_{cell} + P_{gas}$). The temperature difference between the electrolysis cell and the constant-temperature bath is used to determine the heating power of the electrolysis circuit. Once the heat transfer coefficient k is determined the measured power is given by $P_{meas} = k (T_{cell} - T_{bath})$. Calibration and constancy of k, as well as thermal gradients, are key issues of concern for accurate power measurements in systems of this type. In such calorimetric measurements one needs either to know the integral heat transfer coefficient or make appropriate corrections. Fleischmann and Pons employed this type of calorimetry, however, had no mechanical stirring. Isothermal calorimeters are sometimes called isoperibolic or heat leak calorimeters. (Courtesy of Professor Nathan Lewis).

necessary to measure the "heat transfer coefficient" of the calorimeter. This is the most important and crucial factor in calorimetric measurements, a measure of the efficiency of the calorimeter. This coefficient is determined by periodic calibration of the system with an internal heater in the cell, exploiting Newton's law of cooling. On application of power to the internal heater during electrolysis, one traces the temperature-time curve until a steady state is obtained. Then the heater is switched off and the cooling curve is followed. This procedure provides an approximation to the "differential calorimeter constant" at the operating temperature of the cell. More sophisticated procedures are necessary to obtain an integral calibration constant. Once the calibration constant k (in units of watts per degree of temperature rise) has been determined, the output power is given by the product, k ($T_{cell} - T_{bath}$). With an open cell, however, there is still another important correction to the output power associated with the enthalpy of the escaped gases. Eventually this output power is to be compared to the measured input power supplied to the cell, in order to determine if any excess power is present.

As can be suspected already from the above sketchy description of calorimetry, there are a number of more intricate experimental problems in the evaluation of heat effects. One obvious question concerning the determination of T_{cell} is whether the temperature is uniform enough throughout the cell such that the heat sensor (thermometer) will provide a representative measure of it. In the Fleischmann, Pons and Hawkins experiment, no particular precaution was taken to prevent such temperature nonuniformities or gradients. For example, not even mechanical stirring of the cell solution was provided. They relied entirely on the gas bubbling at the electrodes to stir the contents of their cells. The controversy over whether there were temperature gradients in their cells and, therefore, temperature misreadings was addressed by Fleischmann at a meeting where he showed a video of a cell bubbling. On addition of a red dye to the liquid in the cell, it was seen on the video to mix in about 20 seconds. Although this video was presented to show that no temperature gradients existed within the cell, the demonstration would have been more convincing had there been an array of thermometers inserted in various regions of the cell. Caltech experimenters, for example, have reported that mechanical stirring is necessary to eliminate temperature variations in a similar cell.

A number of critics have pointed out the lack of experimental detail in the Fleischmann, Pons and Hawkins' paper. One remarked that it was the first time he had seen a paper on calorimetry where not a single temperature of the cell was given [although the temperature of the constant temperature bath was given in the paper as 300° K (300° Kelvin is equivalent to 27° centigrade), later changed in the errata to 303.15° K]. In addition, no information was given on the crucial heat transfer calibration of the cell.

27

The paper omitted most of the very same essential details that the authors had also constantly avoided to address in their previous contacts with the press. This lack of experimental detail became especially evident when discussion surfaced about the manuscript that the University of Utah had submitted to *Nature* on March 24. The paper submitted to *Nature* was an abbreviated version of the paper that appeared on April 10 in the *Journal of Electroanalytical Chemistry*. The referees for *Nature* had a number of difficulties with the manuscript and requested that the authors answer a number of questions and supply additional critical information before making a decision on publication. John Maddox, Editor of *Nature*, said the article was a shortened version of a previously published paper and, as submitted, would not have told scientists more than they already knew. Deputy Editor Peter A. Newmark said the Fleischmann-Pons' paper "was sufficiently flawed that it didn't pass peer review" (*Business Week*, May 8). Rather than meet the referees' requests, Fleischmann and Pons withdrew their manuscript. This was probably a wise decision[6] since it was highly unlikely that *Nature* would publish an even less detailed version of a similar article that had already appeared in the *Journal of Electroanalytical Chemistry*.

Some additional information about the two different Utah claims reached selected pockets of the scientific community through colloquia presented by Fleischmann, Pons and Jones. For example, Pons spoke at the University of Utah on March 31 and to the Chemistry Department at Indiana University on April 4. Pons started his Utah lecture by complaining about the pressure that he had been under since the March 23 press conference. Pons made another very revealing statement in this seminar, namely, "I have worked in a number of areas of chemistry . . . quite honestly, I would prefer to talk about any one of them now." This appears to imply a certain lack of confidence and an uneasiness in defending their announced claim of watts of heat generated in a simple electrolytic cell. Fleischmann addressed a large audience of European scientists on March 31 at the invitation of the Director-General of the Conseil European pour la Recherche Nucleaire (CERN) in Geneva. CERN is a large forefront laboratory specializing in high-energy particle physics research and operated and financially supported by the European governments. In all of the above lectures, Pons and Fleischmann discussed the material in their article submitted to the *Journal of Electroanalytical Chemistry* but gave no new experimental details. They emphasized the large amount of heat that was released in their experiments, a quantity they claimed to be some 100 times greater than could be accounted for by any chemical reaction. Furthermore, the fusion products (neutrons and tritium) reported were at low levels, some

---

[6] However, a very unusual decision. Most scientists strive to have their prized results published in a prestigious journal.

billions of times less than required to be commensurate with the reported heat generation. This large discrepancy between the heat and fusion products resulted in major skepticism by all those knowledgeable about nuclear physics and penetration of barriers. How could it have been possible to have fusion as the source of heat when the normal fusion products had yields that were too small by factors of a billion? Fleischmann and Pons' answer, that the bulk of the energy was coming from as yet unknown nuclear processes (with, as yet, undetected products), was highly unusual and undefendable to nuclear scientists.

Jones lectured to the Physics Department at Columbia University on March 31. Although he arrived to speak only to a small physics seminar, he was ushered into a large auditorium filled with some four hundred scientists, students and reporters. Based on Jones' presentation, the scientific community began to appreciate that his claim of cold fusion was very different from that of the University of Utah. Jones' fusion process, giving neutrons barely above background neutron measurements, has a rate orders of magnitude too small to be a practical source of energy, but, if confirmed, would be of scientific interest.

On April 12, accounts of the cold fusion experiments at the University of Utah and Brigham Young University were presented at a meeting at the Ettore Majorana Centre for Scientific Culture in Erice, Sicily. A summary of this meeting by R.L. Garwin has been published [*Nature* **338** 616 (1989)]. The work of the University of Utah group was presented by Fleischmann, and that of the Brigham Young group by Jones and J.B. Czirr. The very different claims of the two groups were highlighted at this meeting. Whereas the University of Utah group reported the generation of several watts of power, the BYU group reported only a very small yield of neutrons which, if converted to the equivalent power would be some million million times smaller than the University of Utah figure. Even so, the fusion rate of the BYU group required an enhancement of many orders of magnitude as shown by new theoretical calculations of fusion rates by Steven E. Koonin, theoretical physicst from Caltech. Although these new rates were slightly larger than those from previous calculations, they didn't alter the general theoretical picture that fusion at room temperature is an extremely improbable process. Already at this meeting, less than three weeks after the University of Utah's initial press conference, some groups reported negative results from searches for fusion products from the electrolysis of heavy water with a palladium cathode. Concerning the positive results, many questions remained unanswered due to the lack of experimental details. The last sentence of Garwin's report serves as a summary: "A large heat release from fusion at room temperature would be a multidimensional revolution. I bet against its confirmation."

An interesting personal encounter between Fleischmann and Jones is reported (*Business Week*, May 8, 1989) to have occurred at the Erice meeting. Fleischmann took Jones outside the breakfast room and apologized to him about the events surrounding the March 23 press conference by explaining that there had been enormous pressure from University of Utah administrators to hold the press conference (*ibid.*).

The first claims to have confirmed some aspects of the University of Utah experiments were announced on April 10, eighteen days after the March 23 Utah press conference. These announcements brought a new degree of excitement to the scientific community. In the morning, chemists at Texas A&M University announced results from calorimetric measurements giving excess heat and in the afternoon, physicists at the Georgia Tech Research Institute reported observing the production of neutrons coming from the palladium electrode in a beaker of heavy water. Scientists at both institutions seemed to have been swept up by the momentum of events that were circulating around them. Under such circumstances and in the glare of publicity, several research teams studying cold fusion announced results with uncharacteristic recklessness, even before they had performed obvious checks and had confirmed their own results.

The Texas A&M research team was led by electrochemist Charles Martin. In their news conference the group claimed to have found heat being produced by their electrochemical cell; however, they made no claim to have established cold fusion. The Martin group reported that they were producing 60 to 80% more energy than that supplied to the cell by using a palladium cathode with dimensions of one millimeter diameter and five millimeters length. They claimed that their cell became active after 20 minutes of electrolysis and had been consistently producing energy for three days prior to their news conference.

In contrast to the above claim of heat production, a team of physicists at the Georgia Tech Research Institute, led by James Mahaffey, claimed to have confirmed fusion in an electrolytic cell by the observation of neutrons with a boron trifluoride (BF_3) neutron detector. This group reported 600 neutron counts per hour from their system over a background of forty neutron counts per hour.

Rumors of confirmations were arriving rapidly also from outside the United States. N. Koyama reported at a meeting of the Chemical Society of Japan on April 1 that he had detected large amounts of heat and gamma rays in his cells with palladium cathodes. Others were skeptical of these results since no neutron measurements had been made at this time. Although no details were available, scientists at Lajos Kossuth University in Debrecen, Hungary claimed to have confirmed Fleischmann and Pons results. The electronic mail was humming with rumors of confirmations from several countries including the USSR.

With many rumors and some claimed confirmations of the Fleischmann-

Pons' effect, the atmosphere at the Dallas American Chemical Society (ACS) Meeting (April 9–14, 1989) was highly charged. Dr. Valarie J. Kuck, a researcher at AT&T Bell Laboratories in Murray Hill, NJ, organized a last-minute special session on cold fusion starting at noon on Wednesday, April 12. She had also been responsible for a similar special session on superconductivity at the April 1987 ACS Meeting, shortly after that field became popular.

Invited speakers at the April 12, 1989 ACS session on cold fusion were Harold P. Furth, Director, Princeton University Plasma Physics Laboratory, Allen J. Bard, Professor of Chemistry at the University of Texas, Ernest Yeager, Professor of Chemistry at Case Western Reserve University, B. Stanley Pons, Professor of Chemistry at the University of Utah, and K. Brigitta Whaley, Assistant Professor of Chemistry at the University of California, Berkeley. Clayton Callis, president of the ACS in 1989, opened the session, which came to be referred to by some as the "Woodstock of Chemistry." He excited the seven thousand chemists, gathered in a large arena at the Dallas Convention Center, to an extremely high pitch by his introductory remarks, in which he hailed the tremendous potential of cold fusion as an energy source, and claimed that it might be the discovery of the century. He then went on to detail the many problems physicists were having in achieving controlled nuclear fusion. "Now it appears that chemists may have come to the rescue" he said, and the thousands of chemists in the arena broke into loud applause and laughter. As I witnessed the actions and comments from those around me in the packed auditorium, I could only conclude that a majority of the audience believed they were spectators at a session describing a major breakthrough in energy production. Callis' introductory comments offended physicists and some chemists and polarized the communities of chemists and physicists. The very different reaction of the physicists at the American Physical Society meeting, some three weeks later, is discussed in Chapter VI.

Furth, in an excellent review paper on nuclear fusion, discussed progress toward achievement of practical fusion power. He was the token nuclear physicist speaking at the Dallas ACS session. In his talk, Furth discussed also the extremely small probabilities of fusing hydrogen isotopes at room temperature and the large effective electron mass that would be required to account for the University of Utah claims. Furth concluded that many additional experiments needed to be performed before nuclear physicists would believe the University of Utah's reported data. One of the crucial experiments he suggested was to compare light water (H_2O) and heavy water (D_2O) water under the same electrolytic conditions. Pons replied that he was preparing to do this. On the other hand, based on the discussion following Pons' lecture at Dallas it appeared that Pons and Fleischmann had already performed this control experiment. When Pons was asked why he had not reported results of control experiments with light water substituted for heavy water, he replied "A baseline reaction run with light water is not necessarily a good baseline reaction." When asked to elaborate, Pons intimated he had performed the experiment with light water and had seen fusion, saying "We do not get the expected baseline experiment . . . We do not get the total blank experiment we expected" (*Science* **244**, p. 285).

The next two papers by Bard and Yeager reviewed the science of electrochemistry. Whereas Bard's lecture dealt with electrochemical fundamentals, Yeager's lecture specifically discussed the electrochemistry of the hydrogen-palladium system. Electrochemists were delighted with the attention their discipline was receiving, not only from other chemists, but the larger scientific community. "These have been exciting times for electrochemists", Yeager told the ACS audience. He continued, "I hope your fusion fever stays high." It was surprising that the organizers felt it necessary to have two introductory lectures on electrochemistry, a subject that should be familiar to most chemists and is taught in college freshmen chemistry courses. Might it not have been at least more profitable for this audience of chemists to have had an additional lecture in basic nuclear physics in order for them to understand the problems associated with Fleischmann and Pons' claims.

Pons' lecture at Dallas covered the same material that was published on April 10 in the *Journal of Electroanalytical Chemistry* **261** 301–308 (1989). He emphasized the large amount of energy (up to 50 megajoules of heat) produced in his cells, stating that the energy was orders of magnitude more than could be produced by any known chemical reaction. The Fleischmann-Pons logic went as follows. The energy produced in the cells was so large that the source cannot be chemical in origin. Since the first conclusion was that the energy cannot be due to chemistry, the next step in their logic was that the energy had to be due to nuclear processes. This was postulated even though the intensity of nuclear particles was at least eight orders of magnitude less than the heat. This enormous discrepancy between the heat and particles didn't seem to phase those who believed in cold fusion. Fleischmann and Pons concluded that an unknown nuclear process was occurring! This gigantic discrepancy between heat and particles lies at the heart of the different views of "believers" and "skeptics." Pons in his talk (following the presentation of Furth who had shown a slide of Princeton's complex and mammoth Tokamak)[7] pleased the large audience by

---

[7] The Tokamak Fusion Test Reactor (TFTR) operated by Princeton University is a very large doughnut-shaped machine composed of magnets to confine the hot hydrogen fuel by powerful magnetic fields. Experiments using deuterium fuel alone have produced temperatures of plasmas up to 300 million degrees. However, scientists have not been able to confine the hot plasma for sufficient times and high enough densities to generate more energy than that supplied.

showing his simple cell in a dishpan, that he claimed produced a sustained energy-producing fusion reaction at room temperature. "This is" he dead-panned, "the U–1 Utah Tokamak." The chemists went wild. Pons, in his lecture, made the same serious mistake as was made in the Fleischmann-Pons publication. Based on his interpretation of the Nernst equation (taught in college freshman chemistry courses), Pons concluded that the deuterium pressure in the palladium cathode was equivalent to a hydrostatic pressure of approximately 10^{27} atmospheres! It seems that it was this incorrect conclusion which led Fleischmann and Pons to believe that, in a palladium cathode, the deuterium nuclei would be forced together close enough to fuse. The Nernst equation, applicable under equilibrium conditions, relates the overpotential in an electrochemical cell to deuterium fugacity. A large overpotential does result in a high deuterium fugacity, i.e. a high chemical potential of hydrogen relative to the standard state at one atmosphere, at the palladium surface. However, in the present case it is incorrect to equate the hydrostatic pressure with the fugacity, as Fleischmann and Pons did. This is because of the non-ideality of the deuterium gas and the loss of deuterium by D_2 bubble formation at the cathode surface resulting in a much (many orders of magnitude) decreased equivalent deuterium pressure. Thus, Fleischmann and Pons' primary justification for initiating cold fusion research was flawed from the outset. One of the unanswered questions in the cold fusion saga is how two experienced electrochemists could have initiated their cold fusion research on the basis of such a simple-minded pressure calculation.

The speaker following Pons was B. Whaley of the University of California, Berkeley, who proposed a "boson screening" theory to explain how deuterium nuclei inside a palladium lattice can get close enough together to fuse. Speaking to an audience of mostly chemists, Whaley made some fairly obvious statements about nuclear properties, e.g., the nuclei of deuterium (^2H) are bosons whereas the nuclei of ordinary hydrogen (^1H) and tritium (^3H) are fermions. Then Whaley took the giant step and hypothesized, without detailed calculations, that the energy of repulsion between the deuterium nuclei is largely screened out so that "the particles can get on top of each other despite the repulsion." The reaction of the thousands of chemists in the large arena was one of instant approval in that this theory "supported" the claims of Fleischmann and Pons that deuterium nuclei were fusing in their palladium cathode. As a symptom of the fusion frenzy that characterized this early period, Whaley submitted a manuscript to *Science*, as had many others. In due time, the peer review process pointed out numerous flaws in the paper and it was rejected. A year later Whaley revised her thinking and published a paper [*Phys. Rev.* **B41** 3473 (1990)] reporting that the calculated fusion rate was several orders of magnitude smaller than the experimental rate claimed by Jones.

Following the five formal presentations, the meeting was opened to questions from the audience. Little new information came from this exchange beyond the inconclusive information about the control experiment with light water already mentioned. A number of offers were made to examine the University of Utah palladium cathodes for fusion products. All such potentially important collaborations were refused, probably based on patent infringements. In retrospect, the Dallas ACS Presidential session on April 12 was a most unusual event in the history of science. One of the great mysteries associated with the Dallas ACS meeting was how it was possible for the majority of the seven thousand chemists there to accept the fusion of deuterium at room temperature, when the reported experimental fusion product yields were at least 8 orders of magnitude (10^8) less than the reported heat.

Within a few weeks of the announcement of the cold fusion claims from Utah, 'theories' proposed to explain these unusual reports were propagating faster than a runaway nuclear reaction. This proliferation of so-called 'theories' was one of the greatest surprises in the entire cold fusion episode. In most of these theories there was neither much regard for established scientific principles nor for the accuracy of reported experimental data. The 'boson screening' theory has already been discussed. One of the most publicized 'theories' in the early weeks was due to Peter Hagelstein, an electrical engineer from the Massachusetts Institute of Technology. He apparently believed Fleischmann and Pons' claims despite their very weak experimental evidence. Hagelstein initially proposed that the fusion inside the palladium cathode produced an isotope of helium with mass 4 (^{4}He) and a very large amount of energy that was retained in the metal lattice via collective and coherent effects. He wrote four papers on his work in the weeks immediately following the University of Utah press conference; however, these original papers were predicated on the incorrect postulate that cold fusion was producing helium according to reaction (1c) on p. 6 where the energy was miraculously taken up by the lattice, without producing detectable amounts of the 23.8–MeV gamma ray. Newspaper accounts stated that MIT had applied for patents in competition with the University of Utah. MIT Provost Deutch later acknowledged that the application for patents was a "mistake." Others made similarly erroneous theoretical proposals like that of Hagelstein and one of these theories, the one by Walling and Simons, will be discussed later.

Theoretical physicist George Chapline and two physicist colleagues from the Lawrence Livermore National Laboratory proposed that the small amount of neutron production, as reported by Jones et al., was due to muon-catalyzed fusion. They suggested that sufficient numbers of cosmic-ray muons were available at the Utah altitude to induce a small amount of deuterium fusion in the electrolytic cells. Their paper was submitted to

Physical Review Letters, but was not accepted for publication. Chapline *et al.* underestimated the importance of massive amounts of palladium in cold fusion experiments, where the muons are preferentially captured by the palladium, due to its much larger nuclear charge than deuterium. This was later proved conclusively by a Japanese group who bombarded a palladium cathode loaded with deuterium by a beam of muons produced by an accelerator. Hence, this 'theory' failed even to account for the tiny neutron yield reported by Jones *et al.*, which was some 13 orders of magnitude (10^{13}) less than that commensurate with the amount of heat claimed by Fleischmann and Pons. Peer review by experts in the field worked well in keeping the erroneous paper of Chapline *et al.* from being published.

On April 14, 1989, only three weeks after the University of Utah press conference, Cheves Walling and Jack Simons, Professors of Chemistry at the University of Utah, submitted a 'theoretical paper' to the *Journal of Physical Chemistry*. After some modification this manuscript was published on June 15 [93, pages 4693–4697 (1989)]. Walling and Simons' 'theory' of cold fusion illustrates the depths to which scientific discourse can sink when surrounded by the likes of the fusion frenzy which appears to have saturated certain segments of the Utah population. As described below, this 'theory' violates several well-known aspects of nuclear reactions, requiring the authors to invent a string of marvels or miracles.

Walling and Simons developed their 'theory' to 'explain' the large yield of helium (^{4}He) that was claimed by Fleischmann and Pons to have been observed in mass spectrometer measurements of the gases escaping from their cells. This claim seemed consistent with the dearth of neutrons and tritium found by Fleischmann and Pons. In their manuscript Walling and Simons stated that "the most startling feature of the experimental results of Fleischmann and Pons is that the actual heat production, measured by simple calorimetry, is $10^7 - 10^{10}$ times larger than predicted by the neutron and tritium counts measured." Walling and Simons reported in their paper (based on a private communication from Pons and Hawkins) a ^{4}He production rate of 8 x 10^{12} to 8 x 10^{13} atoms of ^{4}He per second per cubic centimeter of cathode. The missing ^{4}He, the nuclear product that eluded earlier work, had apparently been found. This large production rate of helium exceeded the reported power of 0.5 watt per cubic centimeter of cathode. Walling and Simons, however, dismissed this discrepancy as due to error in the mass spectrometric measurements. Since the claimed ^{4}He yield was so large, Walling and Simons took the reported result at face value, and assigned the major path of energy production to the reaction branch $D+D \rightarrow ^4He+\gamma$ [see reaction (1c) on p. 6]. This was done even though it violated all experimental and theoretical nuclear physics evidence favoring reactions (1a) and (1b) as the dominant D+D fusion channels.

The Walling-Simons 'theory' can best be characterized as a triple-miracle 'theory'.* These authors invoked the first miracle to obtain the Fleisch-mann-Pons reported high rate of power production. To do this Walling and Simons assumed the deuterons in the metal lattice were associated with "heavy electrons" with an effective mass ten times the bare electron mass. Endowing the effective electrons in the lattice with such a hypothetical heavy mass would indeed enhance fusion rates. However, the phenomenon of "heavy electrons" involves long-wavelength excitations in which strong correlations modify the electron wave function near the Fermi surface. Because the proposed nuclear tunneling process in cold fusion must occur at distances orders of magnitude smaller than one lattice site, only the short-wavelength "bare" electron excitations are relevant for screening. Hence, the effective electron mass for the tunneling process is very near the "bare" electron mass and the fusion rate cannot be enhanced significantly by a screening process (U.S. Department of Energy Report DOE/S–0073, November, 1989). Hence, the first miracle was clearly flawed theoretically.

Walling and Simons invoked a second miracle to alter the branching ratios of the three reactions (1a, 1b and 1c on p. 6) that are known to occur when deuterium nuclei fuse. At energies down to a few kiloelectron volts, as is well known, reactions (1a) and (1b) have nearly equal production rates. Furthermore, the third reaction (1c) in which two deuterium nuclei fuse to produce ^{4}He and a 23.8–MeV gamma ray has an extremely small branching ratio of the order of 10^7 times smaller than those of the other two reactions. Approximately the same branching ratios have also been observed for muon-catalyzed fusion,† a type of cold fusion. The Walling-Simons 'theory', by fiat, postulated branching ratios altered by many orders of magnitude, such that reaction (1c) would dominate in the proposed cold fusion of deuterium nuclei. This was done to explain the large ^{4}He yields discussed above (and later retracted as erroneous) and the paucity of neutrons and tritium. This altering of the branching ratios violates all that has been learned about the decay of excited compound nuclei in the last fifty years. When particle channels are open, as in the D+D reaction, the yields of the particle channels (1a and 1b) are expected to be orders of magnitude larger than the radiation channel (1c), even for the claimed cold fusion. This is in accord with Bohr's compound nucleus model. Walling and Simons in alter-ing the branching ratios by many orders of magnitude had to abandon the Bohr model completely.

Walling and Simons invoked a third miracle to make sure that the reaction heat from the D+D→^{4}He+γ process would go entirely into lattice

---

* This is discussed at greater length on pp. 111–113 below.
† This is discussed on page 108.

heat rather than into radiation (the 23.8–MeV gamma ray), which would be detected easily. This naive assumption was imposed in order to maintain consistency with the above miracles one and two. Gamma rays would carry the energy outside the electrolytic cell, and the resulting lethal level of radiation would have been detected easily. So, the authors conveniently eradicated all the detectable gamma rays. As justification they made analogies with the internal conversion process by which the energy of excited electronic states of molecules is converted into heat. However, their analogy to molecules is not valid for nuclear reactions as explained in our DOE/ERAB Panel report (DOE report S–0073, November, 1989):

> Internal conversion allows an atomic electron of an excited nucleus to carry off the reaction energy instead of a photon. This process is understood quantitatively – it is dominant in heavy atoms with tightly bound inner electrons and for low energy (less than 1 MeV) photons. In helium the atomic electrons are loosely bound and the photon is some 23.8 MeV – there can not be any appreciable coupling between the photon and the atomic electrons, and internal conversion or any related process cannot take place at anywhere near the rate that would be required. The assumption of Walling and Simons invokes enhancement of internal conversion by electrons of high effective mass appropriate to the solid; as discussed above, such band structure effects cannot play the role of real high-mass electrons either in screening in sub-atomic distances or in the internal conversion process at MeV energies.

Almost immediately after Walling and Simons submitted their paper, Fleischmann and Pons retracted their claim of ^{4}He production admitting faulty measurements due to atmospheric contamination. (The ^{4}He saga will be described in more detail in Chapter VIII). This admission of error in the helium analysis totally eliminated all justification for publishing the Walling and Simons manuscript with its pyramid of miracles, since no experimental evidence even existed for helium production. It did not, however, preclude publication of the manuscript and subsequent discussions by Walling and Simons of their theory.

Publication of the Walling-Simons three-miracle 'non-theory' is a striking example of the laxity of the peer review process applied to some of the papers during the height of the cold-fusion frenzy. In the present case the editors published some of the reviewers' comments (an unusual procedure) along with the Walling-Simons' paper. These comments were mostly critical. However, two reviewers of the Walling-Simons' paper stated the need to publish the helium data, which were included in the manuscript by way of a private communication from Pons and Hawkins. Justifying the publication of the Walling-Simons paper on the basis that it contained the critical helium data was a serious mistake. At that time in the cold fusion saga, proponents were promoting the search for helium (^{4}He) as the definitive test of the Fleischmann-Pons phenomenon. It was imperative, therefore, to have such important primary data communicated by Fleischmann and Pons themselves in a format that met the standard peer review process. This procedure would have revealed that the ^{4}He data, as reported in the private communication, lacked all credibility and would not have been published. Hence, the motivation for Walling and Simons' 'non-theory' would have disappeared. Among the several so-called theoretical manuscripts on cold fusion, precipitated by the fusion frenzy during the weeks following the University of Utah press conference, the manuscript of Walling and Simons stands out in its outright unconcern for, and neglect of, established concepts in nuclear physics. Invoking by fiat, not one, but three miracles in order to fit highly preliminary and suspicious helium data from the University of Utah group, before waiting to have the data confirmed, is an example of the inappropriate scientific procedure followed so often by cold fusion proponents.

Ten days after the submission of the Walling-Simons paper on cold fusion, a letter by Cheves Walling was published (written, of course, well before the cold fusion story broke) in the Chemical and Engineering News (April 24, 1989) entitled "Fraud in Research." Ironically, this piece makes a number of statements that appear, in some measure, to be directly applicable to the events and circumstances that led to the Walling-Simons publication on cold fusion. In what follows, I list several quotes from Walling:

> The question of fraud and carelessness in research is getting continued publicity . . . There is a growing concern among the public that all is not well, and the probity of the scientific enterprise is coming into question . . . [T]wo major keys to meeting the problem are simply the keeping of good records, and the proper assumption of responsibility by senior investigators. These are matters on the importance of which we all agree, but their implementation is easily eroded by *carelessness, impatience and the rush to get out results* (Italics mine) . . . [T]his side of deliberate fraud, there is, as we all know, a large morass, which I may call subjective treatment of data. Into this we all step, and, occasionally, some of us become badly mired . . . Basically, it involves selecting and arranging data to support a favored hypothesis . . . Since confidence in our own ideas and the ability to present our results in the most favorable light increase with experience, it is here that senior investigators most often go astray . . . [T]he canons of science demand that experiments be clearly described and their results honestly reported . . . [T]he more important the result, the more it is the responsibility of the senior investigator to examine the original data and records, and to check the reasoning by which a crucial conclusion is reached.

What are the lessons to be learned from the events that took place over the three-week period from the University of Utah press conference (March 23) to the submission (April 14) of the Walling-Simons manuscript? Much sage advice has been written in the article quoted in the above paragraph. More specifically, experiments must be repeated a sufficient number of times to make sure the results are free of the obvious systematic errors and are reasonably accurate. This is especially true when the new data directly contradict all the previously available experimental and theoretical evidence, as was the case for the helium claims. Experimenters are responsible for publishing their own results. It was unsatisfactory for important and crucial data to first appear in the literature by authors not answerable for the measurements. Those making the measurements are accountable for the data and have the responsibility to present a full account of all aspects of their measurements. All of these statements are especially true in the case of the helium measurements under discussion, since the confirming case for the Fleischmann-Pons' large excess heat at this time (April, 1989) rested squarely on whether, or not, the helium results were accurate. "The rush to get out results", before the highly suspicious helium results were confirmed, was no doubt a major factor in the early submission of the Walling-Simons manuscript. Another key factor contributing to this episode was the highly charged fusion craze in Utah. Pons said "the Walling-Simons model is consistent with all the experimental data available." The fusion frenzy was highly contagious. Walling stated that "the concept could be the most exciting development of his lifetime, exceeding the Manhattan Project." The importance of confirming the helium yields and making sure they were correct, before going public, appeared to have had low priority. Instead the craving for fame, notoriety and patent rights took precedence over following the normal scientific procedures. As will be discussed later, Walling and Simons are listed on the patents as inventors of cold fusion along with Pons and Fleischmann.

The Walling-Simons 'theory' had many attractive features that appealed to a wide audience far beyond the strong proponents of cold fusion. In fact, the hypothesized mechanism would be a nearly perfect way to do fusion, surpassing in all ways the actual mechanisms of the well-known deuterium fusion reactions. In the hypothesized mechanism, the excited helium nuclei (^{4}He) produced in cold fusion magically transferred its energy to the lattice of palladium atoms, while producing only very small yields of undesirable fusion products such as neutrons, tritium, and gamma rays. Such a mechanism for producing fusion power would be a dream come true, and from this conjectural point of view, it was indeed unfortunate that the helium results were false. There are important lessons to be learned from the way the helium episode was handled.

Soon after having experienced the excitement and publicity associated with their early confirmations of the Fleischmann-Pons effect, two research groups were faced with the embarrassment of having to retract their early claims. James Mahaffey and co-workers at the Georgia Tech Research Institute announced that their neutron detectors, made of boron trifluoride (BF_3), were extremely sensitive to changes in temperature. The "neutron count rate" jumped by putting one's hand near the detector. Hence, their early signal of neutrons was actually due to background. The group of Charles R. Martin at Texas A&M University announced that they were not able to reproduce earlier calorimetric measurements of excess heat. Their earlier positive claims were due to incorrect equipment-related procedures. The forthrightness and honesty of these investigators in retracting their claims was reassuring in the charged atmosphere of the time.

New rumors and announcements of experimental (heat and particles) confirmations of the Fleischmann-Pons effect were springing up from all over the world. However, many of these early experimental confirmations amounted to little since they were quickly followed by retractions. This led to the popular opinion that nothing had been demonstrated except credulousness and sloppy experimental techniques, and served to fuel skepticism. The same was true of 'theoretical' explanations of the Fleischmann-Pons effect, many of which endowed the palladium lattice with magical properties permitting violation of well-known concepts of nuclear physics. As of the middle of April, 1989, the subject of cold fusion was best characterized by the popular short phrase 'fusion confusion'.

IV

A Panel is Appointed

Among those trying to get reliable information on cold fusion were the members of Congress and President Bush. On April 13, a day after Pons' lecture in Dallas, Nobel laureate Glenn T. Seaborg received a telephone call from Washington. He was eating breakfast in a restaurant in his hometown of Lafayette, California, when a wide-eyed waitress told him that he had a telephone call from our nation's capital. Robert O. Hunter, Jr., then Director of the Department of Energy's Office of Energy Research, asked if Seaborg could catch a noon airplane in order to come to Washington and brief White House Chief of Staff John Sununu and President George Bush on cold fusion. Nuclear chemist Seaborg, a senior statesman of science and long time advisor to several presidents, had been the Chairman of the Atomic Energy Commission during the Kennedy and Johnson administrations and the first three years of the Nixon administration.

The following morning (April 14) Hunter and Seaborg went first to the office of Admiral James D. Watkins, Secretary of the Department of Energy, to discuss the cold fusion claim. They then proceeded to the White House for a meeting with Sununu and some of his aides. Seaborg described the University of Utah experiments and indicated that there was a great deal of doubt about the validity of the reported observations. Although cold fusion was given a favorable reception at the Dallas American Chemical Society meeting, nuclear scientists were very skeptical of the Fleischmann-Pons results. The group then went to the Oval office to brief President Bush. During these talks Seaborg informed the President of his and other scientists' skeptical views about the validity of cold fusion. In his opinion, cold fusion needed to be verified by other independent experimental groups. Seaborg suggested that the whole matter be investigated by an expert panel.

Two immediate actions followed Seaborg's trip to Washington. First, Admiral Watkins instructed the ten large national laboratories in the United States, operated and financed by the Department of Energy, to divert immediately some of their major resources to the study of all aspects of cold fusion. A number of research groups in the national laboratories were already immersed in such studies, some dating from the time of the March 23 press conference. However, the directive from the Department of

Energy intensified and focused the research efforts on cold fusion at the national laboratories. Secondly, Admiral Watkins requested the Energy Research Advisory Board (ERAB), a standing committee that advises the Secretary of Energy on all matters of energy policy, to assemble a panel to assess the new research area of cold fusion.

Admiral Watkins gave ERAB the following charge:

(1) To review the experiments and theory of the recent work on cold fusion.

(2) To identify research that should be undertaken to determine, if possible, what physical, chemical, or other processes may be involved.

(3) To finally identify what research and development direction the Department of Energy (DOE) should pursue to fully understand these phenomena and develop the information that could lead to their practical application.

Admiral Watkins requested that the Board provide an interim report on the first item by July 31 and a final report on all items by November 15, 1989.

Appointment of the panel members was the responsibility of ERAB, with suggestions of names coming from many different sources including Seaborg, Hunter, members of ERAB and other professional organizations. My involvement dated from early April when John H. Schoettler, Chairman of ERAB at the time, telephoned me requesting that I serve as chair of the panel. My initial feeling was that the whole cold fusion episode would be short-lived and that it would be wise to delay appointing such a panel. However, the persuasive manner of both Seaborg and Schoettler and the ongoing press reports on cold fusion convinced me that such a panel was necessary and timely from the Department of Energy's point of view for reasons to be discussed in the next chapter. I agreed to serve, provided that a co-chair be appointed to assist in this major task which was to be carried out on the short time frame outlined above. Early efforts were made to convince Mildred S. Dresselhaus, Institute Professor at the Massachusetts Institute of Technology and a former President of the American Physical Society, to serve as co-chair. Although she considered the possibility seriously, her commitments and travel schedule kept her from accepting a position as co-chair. She did, however, agree to serve on the panel and help me contact potential panel members. Professor Norman Ramsey, Higgins Professor of Physics at Harvard, who was contacted next, consented to serve as co-chair of the panel but only during the initial months, through the writing of the Interim Report, because he was scheduled to go abroad in the middle of July for several months.

It was during the National Academy of Sciences (NAS) Meeting in

Washington that the involved task of choosing an expert and well-balanced panel was initiated. Professor Dresselhaus and I took advantage of the NAS setting to identify and contact qualified candidates for the panel. Our prime consideration in selecting panel members was their scientific excellence and reputation. It was essential also, however, to have members with an expertise in a number of different scientific disciplines in order for the panel to evaluate the many different aspects of cold fusion. Since the cold fusion claims were of an interdisciplinary nature, the panel members needed to represent this diversity of fields. This requirement made it obvious that most of the panel members had to come from the scientific community at large rather than exclusively from ERAB. The same is true, incidentally, for most ERAB panels formed to study a particular subject, where the constituency is composed of both ERAB and non-ERAB members. The final membership on the twenty-two person Energy Research Advisory Board Cold Fusion Panel is given in Appendix II.

All of those selected for the panel, except for a single person, agreed to serve. This was remarkable because of the sizeable time commitment associated with membership on the panel and the lack of remuneration for service on such government advisory panels. The success in putting together such a distinguished panel on short notice given the above factors could only be explained by everyone's keen interest in one of the most remarkable scientific claims of the century. The present panel included members with expertise in atomic physics, electrochemistry, geology, materials and solid-state science, nuclear chemistry, nuclear engineering, nuclear physics and theoretical physics. Six of the panel were ERAB members; sixteen were non-ERAB members. The size of the panel was dictated, not only by scientific diversity, but by the tight time restrictions placed on the panel members to produce both their interim and final reports. With the busy schedules and heavy commitments of all panel members, it was unrealistic to expect all panel members to be able to participate in every one of the laboratory visits and panel meetings. Committee members started their work immediately on appointment.

The general discussion at the NAS meeting in April, 1989 was also dominated by the subject of cold fusion. Several lectures addressed the topic as did private discussions that flourished in hallways. Perhaps the most interesting new result in cold fusion that circulated in preprint form at the NAS meeting came from an Italian group led by Professor F. Scaramuzzi of Frascati. This group's research, presented to the NAS audience by Professor E. Amaldi, foreign associate of NAS, opened a completely different area of investigation for the study of the D+D cold fusion reaction. Their new idea was to operate the cell in a non-equilibrium or "dynamic" mode that did not utilize electrolysis (these experiments are sometimes called "dry fusion"). In this work, deuterium gas at sixty atmospheres of pressure was put in contact with fine shavings of titanium in a stainless steel reaction vessel. The temperature of the sample was varied either by heating or cooling. No neutrons were observed during the time the deuterium was absorbed by the titanium, at room temperature or at elevated temperatures. The neutron search was made with a boron trifluoride (BF_3) counter placed near the reaction cell. Such neutron detectors are prone to background problems as was discussed previously in the case of the Mahaffey experiment. When the vessel was cycled to the liquid nitrogen temperature ($-196°$ Centigrade), bursts of counts were reported, supposedly due to neutrons. In addition, the Frascati group claimed another type of neutron emission following the warming of their vessel from liquid nitrogen to room temperature. The count rate they reported followed a bell-shaped curve, rising to a peak of 300 neutrons per ten-minute counting interval, over some five hours. Although many were skeptical of these experimental results, they caused considerable excitement at the NAS meeting and provoked great effort toward verification in many laboratories.

At this juncture in the history of cold fusion, anyone presenting what looked like evidence in favor of cold fusion was given lots of media attention, due to the power of the dream of an inexhaustible energy source; however, those finding no cold fusion effects were reluctant to publicize their findings. Firstly, these groups were cautious, wanting to be sure that they themselves had not made mistakes. Secondly, they suspected there was a secret part of the procedure that Fleischmann and Pons had not revealed. Hence, the early hesitation of groups with null results was understandable. Rumors of null results were beginning to be circulated informally, nonetheless, while scientists were asking themselves how best to present "no effect" results formally. Eventually, strong multidisciplinary groups that were able to do many varied experiments, accompanied by careful controls and cross-checks, were willing to break the barrier and report negative results. Other smaller groups utilized special cold fusion sessions at conferences to report their negative results (see, for example, Chapter VI where cold fusion experiments reported at the Baltimore APS meeting are described).

V

Hearing before a Government Committee[8]

On April 26, 1989 the Committee on Science, Space and Technology of the U.S. House of Representatives held a hearing on recent developments in fusion energy research. The purpose of this committee hearing was to review the dramatic developments that had occurred following the cold fusion claim of Fleischmann and Pons. Only one month had elapsed since the March 23 Utah press conference and the atmosphere at the hearing was highly charged. Congressman Robert A. Roe, Democrat from New Jersey and chairman of the committee, outlined the reasons for the committee hearing. He stated: "We want to understand these startling developments and gain some insight into the potential implications. We feel that we may be standing at the door of a new regime of scientific knowledge." Congressman Robert S. Walker, the Ranking Republican member of the committee, expressed his enthusiasm also by stating "[T]he committee is extremely excited by the opportunity to examine the promise of this new potential source of energy. Cold fusion may eventually turn out to be one of the most significant sources of energy for the world in the 1990s and beyond, and I look forward to hearing from its discoverers and other experts on the implications of this intriguing breakthrough."

The hearings were opened by Congressman Roe. The following remarks quoted here illustrate the high pitch and intensity of the hearings.

> Good morning. In recent weeks, an atmosphere of high excitement and anticipation has permeated the scientific community as startling possibilities for sustained nuclear fusion reactions at room temperature have emerged. The potential implications of a scientific breakthrough that can produce cold fusion are, at the least, spectacular.
>
> At the heart of this excitement is a collaborative experiment conducted by Professor Stanley Pons of the University of Utah, and Professor Martin Fleischmann of the University of Southampton in England. The experimental work took place on the Utah campus and the announcement of results first came on March 23. This announcement preceded the

---

8 The proceedings are entitled *Recent Developments in Fusion Energy Research*, One Hundred First Congress, First Session, April 26, (1989); No. 46.

traditional submission to a scientific journal where the article would be reviewed by other researchers in the field. Since March 23 researchers around the world have attempted to duplicate the experiment of Drs. Pons and Fleischmann with conflicting results.

Our objective in holding this hearing today is to examine the various developments to date, to allow an interchange among experts with differing views, and to help members of the Science, Space and Technology Committee assess the significance of the current information.

The harnessing of fusion energy for eventual commercial use has been an illusive [sic] dream for decades. The United States, as well as other industrial nations, has spent millions of dollars to fund various experimental approaches to generating sustained fusion energy. All of the efforts to date have required both very expensive machines and extraordinary temperature levels. The possibility of creating fusion energy at room temperature was wishful thinking only two short months ago. Today we have gained new hope.

The hope of producing commercial fusion energy is the hope of an energy-hungry world. Energy is the life's blood of mankind's technological society. Energy is a vital ingredient to our national and individual existences. And yet the fundamental determination of where the world's finite energy resources are located was set eons ago for this planet.

The Middle East has over $\frac{1}{2}$ of the world's known oil resources and $\frac{1}{4}$ of global natural gas reserves. The United States has $\frac{1}{4}$ of the world coal reserves, 4% of global oil reserves and 6% of world natural gas. These realities of geography cannot be changed by either power or politics.

What is perhaps most wondrous is that those facts may be superseded by man's intellectual inventiveness and dogged curiosity.

Today we may be poised on the threshold of a new era. It is possible that we may be witnessing the 'Cold Fusion Revolution.' If so, man will be unshackled from his dependence on finite energy resources.

We are extremely pleased to have assembled here the two professors who may have discovered cold fusion and certainly have brought great excitement to the scientific community . . .

The Committee on Science, Space and Technology has a membership of twenty-nine Democrats and nineteen Republicans. In addition to the opening statements by Congressmen Roe and Walker, other Committee members giving statements were Marilyn Lloyd, Chairwoman, Subcommittee on Energy Research and Development, Don Ritter, James Scheuer, Sid Morrison, Ron Packard and Steven Schiff. Their comments were almost uniformly optimistic as represented by the following sentences.

The prospect of having an unlimited, cheap and clean source of energy is very exciting.

Energy is the lifeblood of our Nation, and fusion energy would be an enormous step toward the goal of energy independence. The discovery of cold fusion may turn out to be historic . . .

This cold fusion process also holds the promise for ending our nuclear waste problem.

There was, however, a word of warning from Congressman Scheuer from New York. Indicating that he had done some homework on cold fusion, he remarked to the Chairman "[T]he process so far by which we have learned about this has been more confusion than cold fusion, and there seems to be a feeling about that the process has been more driven by a wish to protect future potential profits than it has been adherence to normal peer review processes . . ."

Representatives Wayne Owens and Howard C. Nielson from the Second and Third districts, respectively, of the State of Utah were present to introduce Pons and Fleischmann, the two witnesses in Panel 1. Their testimony and figures were very similar to the information contained in their published manuscript in the *Journal of Electroanalytical Chemistry* referred to in Chapter III. Pons spoke first. He began his testimony by stating,

[W]hile discussing new research problems with Martin Fleischmann in 1984, as we usually do, the problem of high-energy or high-pressure electrochemical phenomena was considered. We knew that measurements of hydrogen – the concentration behavior of hydrogen which had been placed in two certain metal lattices by electrochemical means indicated that if one were to try to duplicate these processes by hydrostatic means – in other words, hydrostatic pressures – then it was clear that enormous, almost astronomical, pressures would have to be applied. So this indicated to us the possibility of many new areas of research . . . The most intriguing implication was the possibility that under such high energy conditions it indeed might be possible to fuse light atomic nuclei . . ."

Pons in his testimony showed a series of slides, and on one of these slides he displayed the Nernst equation. He used this equation as the theoretical basis for his claim that the confinement pressure of the deuterium in palladium cathodes during electrolysis was approximately 10^{27} atmospheres (when the electrolytic cell had an overpotential of 0.8 volts). Fleischmann and Pons' belief that such enormous hydrostatic pressures of deuterium were attainable in their palladium cathodes led them to conclude that deuterons in the palladium lattice would be squeezed close enough together to fuse. The attainment of such astronomical hydrostatic pressures, however, is incorrect as discussed in our DOE/ERAB Panel's report (*Department of Energy Report* DOE/S–0073). It is true that large overpotentials do result in high deuterium fugacity. However, the non-ideality of the deuterium gas and the loss of deuterium by bubble formation at the cathode surface lead to only modest equivalent deuterium pressure. By comparing cathodic charging of deuterium into palladium with gas charging for a D/Pd ratio of unity, one obtains an equivalent pressure of only 1.5×10^4 atmospheres, a value more than 20 orders of magnitude (10^{20}) less than the Fleischmann-Pons claimed pressure. Hence, Fleischmann and Pons' basic scientific motivation for doing their cold fusion experiments, namely, the attainment of "enormous, almost astronomical pressures", was seriously flawed. At the actual much more moderate pressures, deuterium nuclei do not measurably fuse at room temperature, and hence, the Fleischmann-Pons conjecture that "it indeed might be possible to fuse light atomic nuclei" by D_2O electrolysis producing sizeable amounts of excess heat is incorrect.

Fleischmann's testimony followed that of Pons. He stated, "[Pons] has really given you the essential observations, and what I would just like to do in the few moments I have here before you is to carry out some speculation about the nature of the results and to try and project these results into the future and make some point of comparison with the more conventional approaches to nuclear fusion. . . ." After discussing world-wide high-temperature magnetic confinement experiments and illustrating the plasma parameters on a slide, Fleischmann proceeded to discuss and illustrate the confinement parameters he assigned to his cold fusion cell. As assigned, these confinement parameters (at the appropriate temperatures) were approximately 10^{15} and 10^{36} for magnetic and cold fusion, respectively. Although Fleischmann talked about the astronomical value of his cold fusion confinement parameter, the comparison of the above numbers of the two different processes was meaningless, like comparing apples and oranges. It is of interest that Fleischmann told the committee "Our work was not just a shot into the dark, as people believe. We were guided by reasonable theoretical formulations of what might be taking place." This "reasonable theoretical formulation" was based on the inappropriate use of the Nernst equation which led Fleischmann and Pons to believe mistakenly in gigantic pressures leading to D–D separation distances inside the palladium lattice that were small enough for fusion to occur.

Questions by members of the Committee on Science, Space and Technology were wide ranging and Pons and Fleischmann's answers generally evaded scientific assessment. In response to probing questions about the public release of information prior to journal publication, Pons responded that once a manuscript was accepted for publication, as in their case, a public announcement "is a typical, very typical, thing in chemistry." In my experience, however, such a procedure is rare and not typical in the field of chemistry. In response to the question raised about national laboratories and universities not being able to reproduce the Utah heat results, Pons assured the Committee that his laboratory would do collaborative experiments with the Los Alamos National Laboratory. This collaboration never occurred because of complications due to Utah patent applications. Answers to the several questions about the large discrepancy between the claimed heat and the fusion products were equally unsatisfactory, and sometimes preposterous. Although the excess heat was claimed to be nuclear in origin, the reported fusion product intensity was more than one hundred million times less. Other miscellaneous questions, including those dealing with the magnitude of the needed budgets for research and technological development, were passed to Dr. Chase Peterson, President of the University of Utah. Accompanying Peterson were high powered public relations experts to handle financial questions. Committee members also inquired from Fleischmann and Pons about commercialization of cold fusion and its potential relative to high temperature fusion, uses in weaponry, and the competition from foreign nations. In answering questions about existing hot-fusion programs, Fleischmann is to be credited in his reply that those programs are well founded in theory and well founded in terms of experimental results. In this respect Fleischmann's testimony was not self-serving.

The presence of Pons and Fleischmann at the hearing added a pronounced excitement due to the notoriety these two electrochemists had gained over recent weeks from interviews and major news stories on national TV networks and leading newspapers. It was most striking to have these scientists, by now well recognized, attempting to persuade Congress that the world's future energy needs would emanate from a small glass jar. Based on the general reaction of the Committee members, Pons and Fleischmann's testimony at the hearing appeared to add credibility to their claim with Congress and was exploited in urging the Committee to move aggressively on funding cold fusion research. In order to reassure members of Congress who were still skeptical about the validity of cold fusion, Pons persuasively responded,

[F]or 5½ years I think we have been our most severe critics on that . . .
We have felt sure for 2 or 3 years.

Accompanying Pons and Fleischmann and other members of the University of Utah delegation to the hearing before the House Science, Space and Technology Committee were Dr. Chase N. Peterson, President of the University of Utah and Ira C. Magaziner, President of the consulting firm TELESIS, USA, Inc. This pair comprised the second panel to give testimony on April 26. Magaziner was introduced by Peterson as one of the world's renowned business consultants on issues of world competition.

Magaziner's statement before the Committee was especially tuned to whip up concern about America's future, especially as it relates to and depends on international competition. The Utah team could not have asked for a more articulate public relations spokesperson. He began by stating,

> I am here because I am concerned about my three children and the future prosperity of their generation in America.

Magaziner went on to list a number of inventions by American scientists but where the technology is being exploited largely by Japanese, Korean and European companies. He used as illustrations the following two examples:

> American scientists at RCA invented the color television, but today European and East Asian companies produce over 97% of the world's color televisions including 85% of those bought by Americans. American scientists at AT&T Bell Labs and Texas Instruments invented the base technology that produced the world's first memory chip, but today Japanese companies produce over 80% of the world's memory chips including over 50% of those bought by American companies. And Magaziner's list went on.

The result of all of this, he asserted, is a large negative United States trade balance as foreign countries convert American science into commercial products not only more quickly but with higher quality control.

The most interesting and, to many, persuasive part of Magaziner's testimony was his clever use of statistics from international business competition to paint a doom-and-gloom scenario for the United States if immediate funding were not given for cold fusion research. Using the format of a risk/return analysis, he minimized the loss of "a few million dollars" (in reality, the University of Utah was asking for $25 million up front) in case the Utah science was discredited and he focused instead on both the enormous benefit of immediate funding and the disastrous possibilities awaiting the U.S. if the funding opportunity were passed up.

> [S]uppose that this science is real and it does open up a new energy source in the next decade and becomes a multi-billion dollar or even hundred billion dollar industry in the next few decades. If we dawdle and wait until the science is proven and if we wait for the economists to hold symposia on whether Adam Smith would approve of putting public money into it or if we move cautiously and invest only in basic research or only in defense applications and wait for the spinoff, we will be much slower off the blocks than our Japanese and European competitors, because they won't run the race that way.

Magaziner's business strategy was clear: to convince the Congress to accept the large financial risk associated with proceeding rapidly in support of all phases of the development of cold fusion. "Competitive success is a leading position in a race. If we fall too far behind at the beginning, we may never catch up. The downside risk of that could be hundreds of thousands of high paying jobs for our children, billions of dollars of trade balance and billions in wealth which then will go to someone else . . . The right decision is pretty clear."

Magaziner's final impassioned remarks were meant to display nothing more than an altruistic concern with the future welfare of America.

> So now I hope you can understand why I came here today even though I am not from Utah and have no interest in palladium.[9] I have an interest in America's future. I see this as an opportunity for America both to develop this science into future American prosperity and also to develop a model for how America can regain world preeminence in commercializing other new sciences in the coming decade.
>
> I have come here today to ask you to prevent another TV or VCR or computerized machine tool or solar cell or superconductor story. I have come to ask you to lead so that we will not be the first of our nation's ten generations to leave its children a country less prosperous than the one it inherited. I have come here to ask you, for the sake of my children and all of America's next generation, to have America do it right this time.

Magaziner effectively prodded Congress into believing in the urgency of quick funding in order to remain internationally competitive in this new field. His bottom line was that Congress had to act quickly and could not wait until the science was firmly proved.

Following Magaziner's testimony, Peterson narrowed the focus from America to Utah. He wanted, he said, "to sketch a general context of the science under discussion, to suggest its potential importance for humanity and the planet, and to share with the important committee of the Congress the intellectual and cultural circumstances at Utah that may have played a role in its nurture and expression." After discussing fission and fusion in general terms, Peterson commented that "the beauty of the Pons-Fleischmann experiments lies in their simplicity" and that "the importance of the promise of so-called 'solid-state fusion' is enormous."

Peterson's testimony could have given his audience the impression that the Fleischmann-Pons experiments had already brought the world to the threshold of cheap and unlimited fusion energy. In his testimony he outlined the advantages of fusion, which are well known and apply to nuclear

---

[9] This is in reference to palladium stock which increased sharply in price following the Utah press release.

fusion in general and not only to the claimed Utah process. Committee members could only have been impressed by the many positive attributes of nuclear fusion as compared to the energy generated by nuclear fission and fossil fuel. The problem of nuclear waste is largely eliminated. The cost of fusion fuel (isotopes of hydrogen) is moderate and its availability essentially unlimited. A reduction in the use of fossil fuel will substantially reduce carbon dioxide pollution, which is thought to lead to global warming (greenhouse effect). Furthermore, reducing coal burning and the associated sulphur emission will help to reduce the acid rain problem. If and when nuclear fusion is a practical source of cheap energy, our precious resources of coal, oil and natural gas can be conserved and utilized for producing valuable chemical compounds.

The most interesting aspect of Peterson's testimony was his analysis of what led to the Utah experiments. Here, although not stated explicitly, he hypothesized about why such a "fundamental" discovery occurred in Utah, rather than at one of the eastern "establishment" universities.

> A capacity to see an old problem from new perspectives was required. Chemists, electrochemists, looked at a problem traditionally reserved to physicists. In fact therein lies some of the humor and bite of the scientific controversy that is raging. I would like to think that it may not be by chance that it happened in Utah, at a university which has encouraged unorthodox thinking while being viewed by the world as a conservative, even socially orthodox place. There in fact may be something valuable in isolation from more traditional centers. America has prospered and innovated at the frontier and the University of Utah is still a frontier that attracts faculty who highly value their intellectual freedom.

This Utah fixation on establishment universities is further illustrated by the remarks of David M. Grant, Distinguished Professor of Chemistry at the University of Utah, who called Pons' critics "the mean bullies of the Eastern establishment." (*Time*, May 8, 1989)

It is seldom, if ever, true that it is advantageous in science to move into a new discipline without a thorough foundation in the basics of that field. Scientific breakthroughs are not as easily explained as Peterson asserts. Most scientific advances are the result of collaborative efforts of scientists working at the frontiers of their discipline over a period of time. Rarely has a scientific discipline been moved ahead by someone who lacks a thorough foundation in the discipline. One must have not only a foundation in the discipline but also be aware of the successes and failures in the field and be on the cutting edge of new developments. The reality of how scientific advances occur is in marked contrast to Peterson's testimony where he suggests that there is a scientific advantage to being "in isolation from more traditional centers." Such isolation has too often produced nothing but embarrassment in science. For example, a group working under the direction of Professor Edward Eyring at the University of Utah reported in 1972 that they had created an X-ray laser. The announcement made a big splash at the time, since researchers had tried unsuccessfully for years to develop a laser that used X-rays instead of visible or infrared light. The claim was greeted with considerable skepticism by experts, however, because the researchers were not laser scientists and there was no known physical process that would explain the effect. Even so, several scientists reported theoretical explanations and some produced experimental evidence in support of the claim, while most scientists could not reproduce the effect. Eventually, it was shown that the Utah claim of the discovery of the X-ray laser was erroneous and due to an unaccounted for sporadic effect. Scientists outside Utah referred to this episode as the "Utah effect." The striking similarities with the Fleischmann-Pons claim have led many scientists to wonder whether the "Utah effect" has struck again. [*Science* **244** 420 (1989)]

The discussion following the statements of Magaziner and Peterson dealt principally with the distribution of government funds to research institutions, patents, international business competition and the need to verify the Fleischmann-Pons effect. Mr. Owens, representing Utah's Second District, opened the discussion with unbridled enthusiasm and optimism. He stated:

Some say solid-state fusion may be man's greatest discovery since fire. Others say, as I do, that it may also be the innovation to protect and perpetuate the Earth's dying life support system, more important than the possible salvation of the dying industrial superiority of America. Man cannot stand another century like the last. In those 100 years, we have consumed more of the nonrenewable richness of the Earth than was used during all of man's previous history. We polluted and poisoned our environment with its use, and it literally threatens our continued existence. The revolutionary discovery, solid-state fusion, arrives simultaneously with our entry into the age of true environmental alarm. So, bursting with pride, Utah's Congressional Delegation brings to the committee the prospect of a second economic chance and a second environmental opportunity. This morning we tell you not only of the discovery which may revolutionize the world's energy system, but more importantly, it may be the answer to the preservation of our home, Planet Earth. Within the next two weeks, the United Utah Congressional Delegation will present you with an innovative legislative plan, one which will precipitate a whole new concept for a national partnership for action. It will combine private and public investment and the opportunity for America to develop, engineer and champion the most far-reaching innovation of our time.

These were heady promises! They were especially inappropriate at a time when the highest priority should have been given to verifying the Fleischmann-Pons effect!

Peterson assumed responsibility for convincing the House Science Committee of the need for federal funds and his estimate was as follows: "the figure that comes to mind is twenty-five million dollars from the Federal Government. Maybe that needs to be 125 million dollars someday, but that's of not any importance right now. Twenty-five million dollars would allow us to start the 'onion' growing with State and private sources." Peterson pointed out to the committee that, as an incentive, the state of Utah had appropriated five million dollars to assist in the formation of the proposed center and was looking for support from other sources. "We are prepared to build a novel consortium of federal, corporate, state and university resources if you choose to join us. Without federal participation the race for competitive leadership will be handicapped."

As convincing as Peterson's case for funding may have been to some, not all were persuaded. *The New York Times*, for example, ran a damning editorial on April 30. In commenting on Peterson's request for twenty-five million dollars of federal support is stated:

> Given the present state of evidence for cold fusion, the Government would do better to put the money on a horse. . . . For Mr. Pons and Mr. Fleischmann, the best bet is to disappear into their laboratory and devise a clearly defined, well-understood experiment that others can reproduce. Until they have that, they have nothing. As for the University of Utah, it may now claim credit for the artificial-heart horror show and the cold-fusion circus, two milestones at least in the history of entertainment, if not of science!

The University of Utah's decision to appeal directly to Congress was an attempt to circumvent the conventional federal funding process where preparation and submission of a research proposal is followed by peer review. In the words of committee member Ron Packard, Congressman from California, the peer review process is "incestuous . . . the type of arrangement where the money goes to those that make the decisions." His research indicated that about 60% of federal research funds go to the twenty top universities. The University of Utah, not being a member of this group, was for this reason attempting to circumvent normal procedure. And to assist it in obtaining federal funds for the proposed cold fusion center, the University of Utah retained the firm of Gerald S.J. Cassidy and Associates. This is a Washington lobbying firm noted for helping universities secure funds directly from Congress for projects that have not passed through the peer-review process. The proponents of this procedure of funding feel that is the only way to break the dominance of the well-known research universities.

On the other hand, the opponents of so-called "porkbarrel" science argue that decisions on federal funding must be made on scientific merit and not politics.

The overwhelming importance of patents and hopes for a financial bonanza in the cold fusion episode were not lost on all members of Congress. The following written question was submitted to Peterson.

To what extent is concern over intellectual property rights and patent applications affecting your ability to disseminate technical information on the results of the research work at the University of Utah, and to give access to national laboratories, such as Los Alamos or Oak Ridge, to your scientists and experimental supporters to obtain the details apparently necessary to verify the results under reproducible conditions?

The written answer by Dr. Peterson was a much more straight-forward and honest answer to the patent question than had been given previously in the testimonies of other members of the Utah delegation.

Upon the advice of our patent counsel, it is not possible for the University of Utah to share research results with other laboratories, particularly national laboratories, until the information has been incorporated into a patent application and the application is on file in the patent office. After that, dissemination to others can, and has, been done. This is the usual conflict between science and commercial interest, exacerbated in this case by the potential importance of fusion technology.

Representative Morrison took issue with Peterson's priorities and called for immediate verification. "I think the need to verify is absolutely paramount right now . . . I just would renew an offer made by Pacific Northwest Laboratory to bring equipment at no cost to you and be part of that reverification procedure." Peterson's response was that Pons was in the process of setting up the trade of equipment and people with Los Alamos. It is well known that this collaboration failed to materialize because of Utah's patent restrictions. It is particularly unfortunate that all such early offers of assistance in various aspects of the verification process were rejected by the University of Utah on patent grounds. The precedence of patents and copyright over scientific verification led the University of Utah to isolate itself from the mainstream of science and to seek cold fusion patents far too early.

The third panel of witnesses included Professor Robert Huggins, Department of Materials Science and Engineering, Stanford; Professors Steven E. Jones and Daniel L. Decker, Department of Physics and Astronomy, Brigham Young University; Professor George Miley, Fusion Studies Program, University of Illinois; and Dr. Michael J. Saltmarsh, Fusion Energy Division, Oak Ridge National Laboratory.

Of the above speakers, only Huggins supported the Fleischmann-Pons claim of excess heat. He addressed the issue of whether there is a significant difference when deuterium, rather than hydrogen, is electrochemically inserted into palladium. His reported observation of excess energy using deuterium supported Fleischmann and Pons' contention that a significant amount of thermal energy is evolved when deuterium is inserted into palladium. Although Huggins claimed an excess power of several percent over the applied power in his cells, he made no measurements of nuclear products, and hence, one could not draw any conclusions about the mechanism or source of the excess power. Huggins asserted that unsuccessful experimenters may not have taken into account critical aspects of materials science in performing cold fusion experiments.

Jones and Decker of BYU were introduced by Howard C. Nielson, Representative from the Third District of the State of Utah. As already emphasized in earlier chapters, the BYU result on cold fusion was enormously different from the claim of Fleischmann and Pons. The BYU claim for fusion was based on an extremely small yield of measured neutrons, which, if translated into energy, would be over a million million times smaller than the watt level of energy claimed by Fleischmann and Pons.

One of the major tasks of the BYU panelists was to establish their credibility in fusion research. For example, Jones reported that he had worked on muon-catalyzed fusion since 1981 and, in addition, had initiated research in related areas since joining the BYU physics faculty in 1985. The testimonies of Jones and Decker were of a more scientific nature than those presented by members of the University of Utah delegation. The BYU physicists explained the characteristics of their detector system, such as its sensitivity to detect neutrons of the particular energy expected from a D+D nuclear reaction. In addition, they explained the importance of understanding the sizable cosmic-ray background in deducing the true signal. They also presented a copy of their peer-reviewed paper that was to appear in *Nature* on April 27, 1989. Jones and Decker emphasized that their research results were of scientific interest only at the present time and did not offer a door to commercial energy. Furthermore, their message on funding contrasted sharply with that of their neighboring university. They advocated that all research funding for cold fusion should come through the normal channels of basic research funding, e.g. through the DOE and NSF, after peer review, and that there was no need at this time to establish research centers and special programs by allocating large sums of money to develop cold fusion.

The majority of experimental groups in universities and national laboratories were getting "no effect" results. The activity of these groups was represented by the testimonies of Saltmarsh, Dr. Harold P. Furth (Director, Princeton Plasma Physics Laboratory), and Dr. Ronald G. Ballinger (MIT). These speakers appeared on the program late in the day when most of the committee members had already left the hearing. Hence, the congress persons leaving early did not receive a balanced perspective on cold fusion.

Saltmarsh, reporting for the Oak Ridge National Laboratory, stated that his institution had not observed any excess heat or fusion products. He went on to say, "[I]t is worth emphasizing that, to our knowledge, most of the other institutions with whom we are in contact have a similar status to report. At a meeting last Wednesday with representatives from all of the Department of Energy's major national laboratories, all reported similar efforts and similar results." Saltmarsh pointed out that the vast majority of experiments carried out in the United States and abroad had failed to duplicate the Fleischmann-Pons reported results. He cautioned, however, that the details of these experiments attempting to duplicate the Utah claims had not yet been reported in the scientific literature or at open meetings. He went on to suggest, "[I]t would be most helpful if one or more of the major laboratories were to collaborate with the Utah and BYU groups by bringing a range of different diagnostic equipment to bear on an already working experiment."

Furth and Ballinger were the speakers in panel four. The message from Ballinger was similar to that of Saltmarsh: "As far as the results of attempts by the team at MIT are concerned, we have been thus far unable to scientifically verify any of these results. This is in spite of the fact that we are employing calorimetry and radiation detection methods of even greater sophistication and sensitivity than those of the University of Utah." Ballinger also had some fitting comments and sound advice for scientists and journalists in regard to the importance of the peer review process in scientific work, a stage which had been largely omitted up to this time in the cold fusion episode. He stated:

In the scientific community the soundness of experimental or theoretical research results is evaluated through peer review and duplication. For results such as those reported, whose potential impact on the scientific community and the world are so great, this review process is absolutely essential. Unfortunately, for reasons that are not clear to me, this has not happened in this case – at least so far. The level of detail concerning the experimental procedures, conditions and results necessary for verification of the Pons and Fleischmann results have not been forthcoming. At the same time, almost daily articles in the press, often in conflict with the facts, have raised the public expectations, possibly for naught, that our energy problem has been 'solved'. We have heard the phrase 'too cheap to meter' applied to other forms of electric energy production before. And so the scientific community has been left to attempt to reproduce and verify a potentially major scientific breakthrough while getting its experimental details from the *Wall Street Journal* and other news publications.

Furth's testimony addressed mainly the types of control experiments that would be helpful in proving or disproving the Utah claims. Lacking the answers to the questions he raised, Furth displayed a healthy skepticism and pointed out the silver lining to the current fusion confusion: "[S]ome progress may have been made during the past month in focusing the world's attention on the potential value of a realistic long-term strategy for the achievement of the fusion-energy goal."

In summary, although the cold fusion results had been widely reported upon in the public press, the hearing before the House Committee on *Science*, Space and Technology served to alert the scientific and public communities of the uncertain nature of these claims. The hearing also served to emphasize the importance of distinguishing between the very different results of the two Utah groups. The Utah testimonies were remarkable because of the great difference in claimed power levels. The Fleischmann-Pons report of excess power of the order of magnitude of watts requires fusion rates that are more than 10^{12} (a million million) times greater than those estimated by the BYU group. The much more modest results of the BYU group, if true, are of scientific interest but of no practical significance. On the other hand the results of the University of Utah, should they prove to be correct, would have major implications for future energy production.

The funding scenarios proposed by the two Utah universities differed as markedly as their scientific results. Whereas the University of Utah enlisted the services of high-powered consulting firms to help orchestrate their case for a multi-million dollar Center for Cold Fusion Research, BYU was much more cautious in suggesting that funding for cold fusion be allocated by normal grant procedures with peer review. Many scientists and science administrators found the University of Utah's decision to appear before Congress so quickly with high-powered consulting firms to champion and to request funds for an unproven 'discovery' to be a serious mistake in judgement.

Skepticism about the excess heat claims of Fleischmann and Pons was fueled by several of the panelists from outside Utah who reported that, nationally and internationally, the vast majority of experiments had failed to duplicate the Fleischmann-Pons claims. Even so, skepticism seemed to take a backseat to optimism among the House Committee Members present at the April 26 hearing. It is important to note, however, that a different consensus developed during the weeks after the hearing, namely, that the University of Utah evidence of excess heat was not sufficiently convincing to fund a research center to develop cold fusion. Hence, no federal funds have gone to the state-supported National Cold Fusion Institute at the University of Utah.

VI

Cold Fusion Frenzy Peaks

During the second month, positive and negative results on cold fusion were presented and debated at three large scientific conferences. The Spring meeting of the American Physical Society was held in Baltimore, Maryland during the period May 1–4, 1989. The very next week, starting on May 8, the Electrochemical Society held a meeting in Los Angeles. Although the scientific programs at these two meetings covered a variety of topics, cold fusion papers occupied a prominent place on each program and catalyzed broad public discussion on room temperature fusion in a jar. The third conference hosted by the Los Alamos National Laboratory under the auspices of the Department of Energy was a special workshop devoted entirely to cold fusion phenomena. This meeting was held on May 23–25 at the Sweeney Convention Center in Santa Fe, New Mexico.

In contrast to the believing and uncritical attitude of a sizeable fraction of the audience at the Dallas meeting of the American Chemical Society on April 12, 1989, the audience at the American Physical Society (APS) meeting displayed open hostility and disbelief of Pons and Fleischmann's major claim that four watts of heat were produced in their electrolytic cells for one watt input. W.H. Breckenridge, Professor of Chemistry at the University of Utah, stated that the adversarial relationship between chemists and physicists over cold fusion resulted from "the fact that two chemists have made an unusual discovery with profound implications to an area of physics in which most of the active physicists never talk to chemists" (Letter to Editor of the *Wall Street Journal*, April 26, 1989). Although this statement displayed Breckenridge's confidence in his colleagues at the University of Utah, it missed the heart of the controversy, where the reported magnitude of the fusion products was more than 8 orders of magnitude (10^8) less than the reported magnitude of the heat. All critical scientists, including both chemists and physicists, realized the importance of this very large and damning inconsistency.

The sessions devoted to research on cold fusion in Baltimore were held on the evenings of May 1 and 2. The first session started at 7:30 p.m. with some 1800 people present and continued to 12:30 a.m. with several hundred people still in attendance. Some of the principal speakers were S.E.

Jones[10] (BYU), J. Rafelski (University of Arizona), S.E. Koonin (California Institute of Technology), N. Lewis (California Institute of Technology), M. Gai (Yale University) and W. Meyerhof (Stanford). In addition, there were several short contributed papers. In all, some 40 abstracts were submitted for this special session on cold fusion. Due to the lateness of the hour and the number of remaining research groups wanting to report their results, the session on cold fusion was continued on the evening of May 2. The first talk by D.R.O. Morrison (CERN) was a survey in which he presented some disturbing statistical evidence showing a significant correlation of reported negative and positive results with specific regions of the world. For example, in North-Western Europe and the Eastern and Western sections of the United States, all groups had reported negative results except for a single positive result by the group of Huggins *et al.* in the Department of Materials Science and Engineering at Stanford. The groups in these parts of the world generally had lots of sophisticated equipment and experienced staff to set up and execute experiments quickly. In other regions of the world where research groups often had less equipment available, especially for cross-disciplinary experiments, only those few laboratories which had apparent confirmations had reported their results.

Noticeably absent from the Baltimore APS program were Pons and Fleischmann. Fleischmann had been invited to give one of the plenary lectures; however, he had declined, no doubt anticipating a hostile audience. This opened up a position on the program for chemist Nathan Lewis of Caltech to describe his electrolysis experiments to the physicists.

Lewis' lecture made a major impact on the audience and served as a lightning rod to shift the general public's opinion on cold fusion from positive to negative. Lewis reported for a very large collaboration of fifteen chemists and physicists working on cold fusion at Caltech. Their range of expertise was broad and complementary and included the pertinent areas of electrochemistry and nuclear-particle detection. The diversity of backgrounds of the Caltech collaboration made this one of the strongest groups among those performing cold fusion experiments.

The Caltech group was one of the first groups in the world to attempt to reproduce the Fleischmann-Pons effect. Professor Charles Barnes of the Kellogg Laboratory of Caltech and co-leader of the Caltech collaboration, had, by chance, heard of the Fleischmann-Pons claims prior to the Utah press release. A visitor at the laboratory had heard a foreign radio station quoting the *Financial Times* of London (see Chapter I explaining how the

---

[10] Jones had been invited to give a paper at the APS meeting some months ago and this no doubt had contributed to the decision by the University of Utah to submit their paper for publication on March 11 and to proceed with their press conference on March 23, 1989.

Figure 4. Members of the California Institute of Technology collaboration with a cold fusion experiment in the chemistry laboratory of Professor Nathan S. Lewis. Left to right (back row) are Amit Kumar, Reginald M. Penner, Michael Heben and Eric Kelson; left to right (front row) are Lewis and Sharon Lunt. Lewis' lecture on May 1, 1989 at the Baltimore American Physical Society Meeting played a role in shifting public opinion on cold fusion from positive to negative. It is ironic that Lewis was invited to give a plenary lecture on electrolysis experiments when Fleischmann declined his invitation.
(Courtesy of the Caltech magazine *Engineering and Science*.)

Times received the story a day early) which told of a new, boundless source of energy resulting from nuclear fusion at room temperature.

The Lewis-Barnes team embarked immediately on a large and detailed series of experiments that employed a variety of metallurgical pre-treatment procedures of the palladium cathodes and, in addition, used several different electrolytes. At this early stage in the history of cold fusion, it was essential to have done these electrolysis experiments under a large variety of conditions including size and pre-treatment (e.g. cold-worked and cast palladium) of the cathodes, current levels, charging times, etc. Lewis' explicit and thorough account of their enthalpy measurements from this diversity of cells operated in both D_2O and H_2O electrolytes, made a strong impression on the Baltimore audience. After correcting for the enthalpy of the D_2 (or H_2) and O_2 gases escaping from the system, the team obtained agreement between the power input in the palladium-platinum circuit and the power production in the electrolysis cell to better than 6%. They considered this good agreement, basing their expectations on conventional chemical reaction processes for the electrolysis of water. In addition, this result was far below the excess heat claimed by Fleischmann and Pons.

The Caltech group also looked at their operating electrolytic cells for production of neutrons, tritium, helium and gamma rays (these are the products of the fusion reactions shown on page 6 as reactions 1a,1b, and 1c). In none of the electrolysis cells did the Caltech group observe a statistically significant neutron count that exceeded the background level, a result in sharp contrast to the initial Fleischmann-Pons claims. The neutron detector used in these experiments was developed at the Kellogg Laboratory for studies in astrophysics. It consists of a 40 centimeter cube of polyethylene with a 10 by 10 centimeter horizontal channel through the center in which the working electrolysis cells were placed. Neutrons were thermalized in the polyethylene and detected with twelve ^3He proportional counters that were embedded in the cube around the central channel. The cube was surrounded by various arrays of shielding materials to reduce backgrounds. In addition, software cuts in pulse height spectra and timing spectra further reduced the background. They obtained an upper limit on the difference between the foreground and background rate of 100 neutrons per hour at the 95% confidence level. Based on a 0.22 centimeter diameter by 10 centimeter long palladium (Pd) cathode, this corresponds to an *upper limit* of 0.070 neutrons per second per cubic centimeter of Pd. This upper limit is to be contrasted to the neutron flux of 3.2×10^4 neutrons per second per cubic centimeter of Pd reported (40,000 neutrons per second for a Pd cathode of size 0.4 by 10 centimeters) by the University of Utah group [*J. Electroanal. Chem.* **261** 301 (1989)]. Hence, the Caltech upper limit on the neutron flux is more than 5 orders of magnitude (10^5) smaller than the neutron flux reported by Fleischmann, Pons and Hawkins!

Before proceeding to discuss the limits established by the Caltech group for other nuclear particles, some further comment is in order about the neutron measurements made by Fleischmann, et al., which were also discussed at the APS meeting. Rather than count the neutrons from the D+D → ^{3}He + n reaction directly, the Fleischmann et al. procedure relied on the thermalization and capture of fusion neutrons by hydrogen [H + n (thermal) → D + γ(2.224 MeV)] in the water bath surrounding the cell. The resulting 2.224–MeV gamma rays, detected with a 3-inch diameter by 3-inch height sodium iodide crystal scintillation detector, served as the signal for fusion neutrons. This method for detection of the fusion neutrons is several orders of magnitude less sensitive than the direct counting of these neutrons as was done by the Caltech group.

The validity of the Fleischmann et al. neutron results suffered a major blow from the detailed measurements and analyses by R.D. Petrasso and collaborators of the Massachusetts Institute of Technology (MIT), which were presented at the APS meeting and later published in *Nature* 339 183 (1989). These authors studied neutron capture by hydrogen using a known plutonium-beryllium neutron source submerged in water and concluded that the claim of Fleischmann et al. [*J. Electroanal. Chem.* 263 187 (1989)] to have observed the 2.224–MeV gamma ray line from neutron capture on hydrogen was unfounded. The MIT investigators based their conclusion on three quantitative considerations. First, the linewidth of the gamma-ray line reported by Fleischmann et al. was a factor of two smaller than their instrumental resolution would allow; secondly, the Fleischmann et al. gamma-ray spectrum did not contain a Compton edge at 1.99 MeV, which must be present with a 2.224–MeV gamma ray due to a scattering process between the photon and an atomic electron; and thirdly, their estimated neutron production rate of approximately 4×10^4 neutrons per second was too large by a factor of fifty when compared to the intensity of their reported 2.2–MeV gamma-ray production rate. Based on these three pieces of experimental evidence, the Fleischmann and Pons signal line cannot be the 2.224–MeV gamma-ray line originating from thermal neutron capture on hydrogen. Fleischmann et al. had already admitted that their heat generation was more than eight orders of magnitude larger than their reported neutron flux. They had explained this gigantic discrepancy by stating that the heat generation had occurred through a hitherto unknown nuclear-fusion process. Now Petrasso et al. showed that Fleischmann and Pons' reported signal line was due to an instrumental artifact rather than a 2.224–MeV gamma ray! This finding substantially weakened the Fleischmann et al. claim that nuclear fusion was occurring in their cells since no evidence for the detection of neutrons now existed. The controversy surrounding the Fleischmann-Pons claimed evidence for the detection of fusion-produced neutrons is discussed in Chapter VIII.

Within their experimental detection limits, the Caltech group also found no evidence for nuclear-fusion-produced tritium, helium and gamma rays. The electrolyte in their cells was analyzed for excess tritium that might be produced by reaction (1b) on page 6. These measurements were performed on a Beckman Instruments LS–5000 TD scintillation counter, and no excess tritium was observed. The ^4He in the evolved gases at palladium cathodes reported by Pons et al. on television and in the Walling-Simons preprint prompted the Caltech group to investigate the composition of the gases in the headspace of their electrolysis cells. The analysis for helium was performed with a VG model 7070E magnetic-sector mass spectrometer, the same model instrument as was used by the University of Utah group. The search for high-energy gamma rays was made with large intrinsic-germanium and sodium iodide detectors. Conservative limits were placed on the intensities of the 5.5-[H+D → ^3He + γ(5.5 MeV)] and 23.8–MeV (reaction 1c on p. 6) gamma rays. The high-resolution germanium gamma-ray spectrum also showed no evidence for gamma-ray lines arising from Coulomb excitation of palladium isotopes by charged particles associated with fusion.

In addition to reporting the results of the Caltech collaboration, Lewis also reported on his analysis of the published calorimetric data of Fleischmann et al. [The Caltech group later published these results in Science 246 793 (1989)]. Lewis gave estimates of the raw data that are the basis for Fleischmann and Pons' claims of large excess power production. This analysis was particularly revealing since the popular press had repeatedly stated that the Fleischmann et al. cells yielded a power output of 4 watts for 1 watt of input power, giving an excess power [relative to the input power, i.e. (4 watts − 1 watt)/(1 watt)] of 300%. Such large power excesses were hypothetical numbers given in column c of the paper by Fleischmann et al. [J. Electroanal. Chem. 261 301 (1989)]. In contrast, the actual average excess power for nine cells was 27% (column b). Cylindrical cathodes of dimensions 0.1 by 10 cm, 0.2 by 10 cm and 0.3 x 10 cm were each run at current densities of 8,64 and 512 milliamperes per square centimeter. It is important to note that the reported values of the excess power, relative to the values of the input power, were *very much smaller* than the values that had been frequently reported in the public press.[11] In view of these much more

[11] The source of all the statements about the University of Utah's large power gain of 300 percent originated from the March 23, 1989 press conference. This was an extrapolated number and not a *measured* number. This much exaggerated number was propagated in an article written by Jerry E. Bishop (*Wall Street Journal*, March 24, 1989). He stated, "An experiment now in operation is producing four watts of power for each watt of input, said James Brophy, vice president of research at the university." In Bishop's next article (*Wall Street Journal*, March 27, 1989) he said "It was the assertion last Thursday [i.e. March 23] that the Utah fusion device was producing four watts of energy for each watt put into the device that astonished

modest values of the actual excess power reported in the first publication of Fleischmann and Pons, Lewis emphasized the importance of making accurate error analyses on each of the experimental quantities and the necessity of making a critical evaluation of each of the assumptions which are required to transform the raw data to excess power production.

Lewis' lecture was devastating to the credibility of the Fleischmann et al. claims. From their very extensive set of measurements, the Caltech group had found no evidence for low-temperature nuclear fusion of deuterium in palladium. They reported neither excess power from their calorimetric measurements nor nuclear products from detailed particle and gamma-ray searches. Based on neutron counting, an upper limit, at the 95% confidence level, of 1.5×10^{-24} $D(D,n)^3He$ fusions per D–D pair per second was established. A report of this work was later peer-reviewed and eventually published in *Nature* **340** 525 (1989).

The calorimetric power measurements that were carried out by the group of R.A. Huggins of Stanford University were criticized by W.E. Meyerhof in his lecture at the APS meeting. Huggins et al. were claiming to have confirmed the Fleischmann and Pons work by observing excess power, although they were not claiming that this power was due to nuclear fusion. Meyerhof, along with D.L. Huestis and D.C. Lorents of the Stanford Research Institute (SRI) made calculations of the heat flow and temperature gradients in Huggins' cells. Meyerhof reported that power excesses were sensitive to the arrangement and positions of the components in the cell and concluded that the claimed energy generation of Huggins et al. and Fleischmann et al. were due to poor calorimetry and inadequate accounting of their data. Although Meyerhof had tried very hard to review his calculations with Huggins, he was not able to do so before his APS talk. Three weeks later at Santa Fe, Huggins reported that they had resolved some of their differences on the interpretation of his calorimetric data, however, major open questions remained, and Huggins' results were still in question.

The theoretical talk by S.E. Koonin emphasized the extremely small probability for the fusion of hydrogen isotopes at room temperature. He presented new calculations of cold fusion rates in diatomic hydrogen molecules involving the different isotopes. The rate of D+D fusion was calculated to be 3×10^{-64} per second, some 10 orders of magnitude (10^{10}) faster than a previous estimate. Even so, this new rate is extremely small when compared to the claimed rates of either Fleischmann et al. or Jones et al. In fact Koonin's new theoretical rate is still so small that it corresponds to only

many physicists . . . Professors Pons and Fleischmann . . . had announced at a Salt Lake City news conference that they had achieved a sustained hydrogen fusion reaction at room temperature. The reaction was producing more energy than it consumed, they said, citing the four watt figure."

one fusion occurring per year for a mass of deuterium equivalent to the mass of the sun! The theoretical fusion rate of the p+D reaction was calculated to be 10^{-55} per second, over 8 orders of magnitude (10^8) faster than the D+D reaction. This increased rate (although it is still extraordinarily slow) is due to the enhanced tunneling in the lighter system.

In contrast to the audience of largely believers at the American Chemical Society meeting in Dallas, which had responded by cheering when ACS president Clayton F. Callis indicated that the decades-old problem of getting energy from fusion may have been solved by Pons and Fleischmann, the audience of largely skeptics at the American Physical Society applauded loudly when they heard Koonin explicitly state that the pair's results were wrong. Koonin expressed his opinion of the pair in no uncertain terms: "We're suffering from the incompetence and perhaps delusion of Drs. Pons and Fleischmann." Many other doubters at the APS meeting were equally outspoken. Eight of nine speakers polled at the APS press conference went on record to say that they were confident that the University of Utah's claimed heat excess was not due to nuclear fusion. The lone dissenter was Johann Rafelski of the University of Arizona. Although he agreed that the credibility of Fleischmann and Pons was "dented", he indicated that the controversy had not yet been settled because Fleischmann and Pons were not present to respond to the criticisms.

When Jones of BYU stepped on stage the focus was off Fleischmann and Pons' preposterous claim and onto Jones' own claim of cold fusion, which however, bore almost no resemblance to the Fleischmann and Pons' claim. Jones had been invited to speak at the Baltimore meeting in advance of the University of Utah press release. The Jones et al. paper submitted to *Nature* on March 24, 1989 appeared on April 27, prior to the Baltimore meeting. Jones' claim differed in two important respects from that of Fleischmann and Pons. Aside from not making calorimetric measurements, Jones *et al.* reported a neutron intensity from their electrolysis experiments that was more than 5 orders of magnitude (10^5) below the neutron intensity reported by Fleischmann and Pons. Of greater significance, Jones' neutron intensity was 13 orders of magnitude (10^{13}) smaller than that required to produce the excess heat of Fleischmann and Pons (if one assumes the known branching ratios for the D+D reaction). Even so, the Jones *et al.* result of 1×10^{-23} neutrons per DD pair per second, assuming fusion is a volume effect, is a startling one. (This rate was reported initially for a single run; later when averaging over several runs, the rate decreased by a factor of six.) As was mentioned earlier, Koonin calculated the fusion rate for a pair of deuterium atoms in a diatomic molecule to be 3×10^{-64} per second. Jones' claimed enhancement of over a factor of 10^{40} (40 orders of magnitude) when the deuterium occupies lattice sites in the cathode during electrolysis is an enormous effect, and of great scientific interest, if true. Even though the

Jones *et al.* claim is revolutionary, it in no way compares to the Fleischmann and Pons' claim, since no excess heat is claimed and the process has no practical application as a useful energy source.

Since Jones and his collaborators suspected that the fusion rate would be extremely small, they developed a neutron spectrometer that was sensitive to the identification of 2.45–MeV neutrons (see reaction (1a) on p. 6), and that minimized the cosmic-ray background. The Jones *et al.* total neutron counting rate was very near the neutron background, and established the necessity of having an accurate knowledge of the background. Hence, it was understandable that only one experimental paper at the Baltimore APS meeting challenged their claim of a net positive neutron count. All of the other experimental papers at this time reported upper limits on the neutron intensity that were of the same order of magnitude or larger than the Jones *et al.* value, but orders of magnitude smaller than that reported by Fleischmann and Pons. Some of these researchers reporting null results in terms of upper limits on the neutron intensity were from IBM's Thomas J. Watson Research Center, AT&T Bell Laboratories, Oak Ridge National Laboratory, Lawrence Berkeley Laboratory, Chalk River Nuclear Laboratory of Canada, Florida State University, University of Rochester and the University of Toronto.

Moshe Gai, the spokesman for the Yale-Brookhaven National Laboratory collaboration, gave an impressive account of their experiments at Baltimore, where their upper limit on neutrons was considerably below the Jones *et al.* result. This group measured neutron and gamma-ray emission from a variety of electrochemical cells using a sensitive detection system with a very low background. No statistically significant deviation in counting rate from background was observed in either the neutron or gamma-ray detectors. The Yale-Brookhaven upper limit on the neutron rate was 2×10^{-25} neutrons per deuteron pair per second, more than an order of magnitude (10^1) less than the Jones *et al.* result and some 6 orders of magnitude (10^6) less than the result reported by Fleischmann and Pons. The detector system of the Yale-Brookhaven National Laboratory group was state-of-the-art with a reasonable efficiency for neutron detection and a very low neutron background. Hence, this group was one of the first groups to cast doubt on the claim of Jones *et al.*

At the APS press conference, the above mentioned group of nine physicists also voted on whether or not they thought the production of neutrons at the very low level reported by Jones *et al.* was due to nuclear fusion. Three of the nine voted that the neutrons were not associated with nuclear fusion. This is to be compared to the eight who asserted that Fleischmann and Pons' reported heat results weren't due to nuclear fusion. As suggested by the straw votes of the two groups, the majority of those physicists present in Baltimore accepted the possibility that Jones' reported results were correct [even though Jones' results represented a startling enhancement in the theoretical fusion rate of 40 orders of magnitude (10^{40})] while rejecting the heat and particle measurements of Fleischmann and Pons as flawed and simply incorrect. Except for Jones' neutron results, all of the particle measurements reported in Baltimore, both by those giving invited and contributed papers, were negative, and established upper limits for fusion neutrons that are orders of magnitude below the reported values of Fleischmann and Pons.

A few of the theoretical papers at Baltimore were positive, and could be characterized as attempts to 'explain' how the phenomena might occur. Rafelski's abstract, for example, explained nuclear fusion at the Jones' level: ". . . [T]he fusion rates observed by the BYU team of S.E. Jones during electrolytic infusion of hydrogen (D) into Pd and Ti cathodes can readily be explained by a combination of standard nuclear physics data and WKB penetration integrals in the metal lattice environment." He later retracted this statement. In a much more striking abstract, Michael Danos of the National Institute of Standards and Technology produced a theory that was able to 'explain' the Fleischmann and Pons reported results. "The order of magnitude of the resulting [theoretical] rates corresponds to the observed rates of 10^{-10} sec^{-1}." He also had a mechanism for explaining the suppression of protons, neutrons and photons. Not only did his claimed fusion rate agree with the heat claim of Fleischmann and Pons, the suppression of nuclear particles was 'explained' also. Danos is reported to have said "In my view, there is no doubt that direct fusion does take place and Pons may be seeing it" (*Business Week*, May 8, 1989). A few months later Danos came to a completely different conclusion and retracted his 'theory' also.

The mood at the Baltimore APS meeting was very different from that at the Dallas ACS meeting which had occurred three weeks earlier. The large number of negative papers at Baltimore, which reported upper limits on fusion-particle intensities that were orders of magnitude smaller than those claimed by the University of Utah group, served to swing the pendulum of opinion towards that of the skeptics. In a sense the fusion frenzy had reached a high water mark during the Baltimore APS meeting. On the following Monday (May 8, 1989), several national news magazines, including *Business Week*, *Newsweek*, and *Time*, ran cover stories on cold fusion. It is most unusual for the latter three magazines to run the same cover story on the same week. Hence, the occurrence of the same science story in all three magazines illustrates the public's early interest in room temperature fusion triggered by the University of Utah's announcement. Media reports based on the papers at the Baltimore meeting, however, served to dampen the public's optimism that cold fusion would produce useful energy. This was in sharp contrast to early reports, e.g. in the *Wall Street Journal*, where the cold fusion process was hailed as one of science's greatest gifts to mankind, promising a golden age of cheap, clean and nearly inexhaustible energy. Most scientists were critical both of what Fleischmann and Pons claimed and what they did not claim. First, physicists said they did not see how the claim of large amounts of excess heat from fusion at room temperature could be possible, and secondly, cold fusion was impossible because there was no evidence of commensurate levels of the by-products that are a crucial part of the signature of a fusion reaction. Cold fusion had become a profound frustration for many scientists, who reported, initially with some embarrassment, that they were not able to reproduce the Fleischmann-Pons' experiments. With the negative reports from mainstream scientists at Baltimore, the consensus of those attending the meeting was reasonably well described in the following words by Koonin: "I find the BYU results theoretically dubious but not impossible. I find the University of Utah results theoretically impossible."

Following the Baltimore APS meeting, the University of Utah experienced two major setbacks in their quest to obtain federal support for their cold fusion experiments. On the night of May 3, 1989, Fleischmann and Pons went to Washington, D.C. for a meeting the next afternoon with John Sununu, President Bush's Chief of Staff. On the morning of May 4 the Utah scientists were notified by the White House that the meeting had been cancelled. Whether or not the negative publicity that came from the physicists at the Baltimore APS meeting had played a role in cancelling the scheduled meeting with Sununu cannot be answered conclusively; however, the circumstantial evidence would suggest it. The same can be said for the subsequent cancellation of a meeting scheduled in Salt Lake City for the weekend of May 13. Utah Representative Wayne Owens had arranged a demonstration of cold fusion at the University of Utah for several key members of the Space, Science and Technology Committee. Owens had been working on a bill to create a national nuclear-fusion center at the University of Utah and was anxious to bring the national legislators to Salt Lake City for an on-campus lobbying session. Hence, the cancellation of the trip by Representative Roe, Chair of the Committee, was a severe blow to the University of Utah (see Chapter V for a discussion of the Hearing before the Committee on Science, Space and Technology on April 26).

On May 8, 1989, exactly one week after the meeting of the American Physical Society in Baltimore, some 1500 chemists jammed into a special night session on cold fusion in Los Angeles sponsored by the Electrochemical Society. This group was expected to be considerably kinder and more sympathetic to the Utah chemists Pons and Fleischmann. Several more people were expected to present confirmations, and Pons had promised that he would give a more detailed explanation of the group's heat measurements. At the previous meeting it was this lack of information that had caused the physicists in Baltimore to attack the University of Utah experiments as faulty and lacking evidence of fusion.

Since Fleischmann and Pons are card-carrying electrochemists, it was appropriate that both chose to attend their society's meeting in Los Angeles. Their presence caused excitement and expectation among the audience. During their formal presentations, Fleischmann and Pons attempted to respond to some of the criticisms raised in Baltimore. First, they tried to answer charges of critics who suggested that there were temperature gradients in their cells. To do this, Fleischmann showed a videotape of the addition of a red dye to their electrochemical cell. The dye was mixed in the cell by the bubbling gases in approximately twenty seconds, indicating, according to Pons and Fleischmann, that the temperature in the cell was quickly brought to uniformity. This video, however, did not satisfy critics such as Nathan Lewis, who had previously made careful measurements of the actual temperatures throughout a similar cell and had demonstrated that temperature gradients did exist when no mechanical stirring was done. Hence, a more prudent experimental technique would have utilized stirring along with direct measurements of the temperature at various locations in the cell. In answer to the criticism raised in Baltimore that their experiments hadn't produced any excess heat, Fleischmann and Pons reported for the first time the surprising claim that one of the Utah cells had given off a burst of heat lasting up to two days. They claimed that the heat produced during this unexpected burst was some fifty times the electrical energy put into the cell. In addition, they claimed that this large energy burst was nuclear in origin because the energy was much larger than that which could be produced by electrochemical reactions. If the energy burst were truly nuclear in origin, equivalent amounts of fusion products must have been released. Without this confirming evidence of commensurate yields of fusion products, Fleischmann and Pons' claim did not convince most scientists that a nuclear process had occurred.

At the Los Angeles meeting, confirmations of excess heat came from Huggins at Stanford, Bockris and Appleby at Texas A&M and Landau at Case Western Reserve. These researchers all reported different amounts of excess heat and different criteria for success. In attempts to produce excess heat, Bockris emphasized that there was "considerable irreproducibility" while Pons stated that he could now ". . . reproduce it more than 90% of the time" (*Salt Lake Tribune*, May 10, 1989). Although these two advocates of cold fusion disagreed sharply on the question of reproducibility, they, along with others, claimed success in producing excess heat. It was surprising, therefore, that a number of large research groups, comprising both chemists and physicists, with sophisticated instrumentation had been unable to observe either excess heat or fusion products. Among the respected institutions where groups at this time were reporting null results were several U.S.

Department of Energy National Laboratories, Caltech, MIT, and the United Kingdom Atomic Energy Authority's Harwell Research Laboratory. What was initially described as an experiment that anyone could perform in a simple freshman chemistry apparatus had turned out to be so complicated and demanding that most first-rate research groups could not reproduce it.

The negative results reported by Lewis made him a target for criticism, most of which was superficial. Stanford's Huggins remarked "[T]he University of Utah's critics, like Nathan Lewis, are exposing their own shortcomings" (*Salt Lake Tribune*, May 10, 1989). Huggins continued, "It's hard for him (Lewis) to see the effects because he is doing it wrong . . . The key is making sure that the crystalline structure of the palladium is such that the deuterium atoms can be loaded inside." Lewis countered, "There's no question that our rods (i.e. the palladium cathodes) were loaded. There's just no fusion going on." Like Fleischmann and Pons, Huggins, with no patents at stake, also refused to share his "secrets" for reproduction with other scientists such as Lewis. It is noteworthy that Lewis, at this time the principal and most vocal critic of the University of Utah experiments, is not a physicist but an electrochemist, as are Fleischmann and Pons, and as such, he was intellectually at home at the Los Angeles meeting on electrochemistry. Here he described again, as he had done the previous week in Baltimore, the large variety of experiments performed by the Caltech team. In none of these experiments had they detected any evidence of cold fusion, neither excess heat nor fusion products including neutrons, helium, tritium and gamma rays. An interesting footnote to the Los Angeles conference was that Lewis had a hard time getting on the program. The organizers of the special session had initially only invited contributions from research groups with positive results including, of course, Fleischmann and Pons, and had not invited groups with negative results. One of the organizers explained with perfect logic that "since the subject of the session is fusion, papers that don't report seeing fusion would not be relevant" (R.L. Park, May 6, 1989). This unacceptable stacking of the deck in favor of cold fusion backfired, leaving many of the electrochemists at the meeting joining the ranks of the skeptics. The press conference at the Los Angeles meeting brought Lewis, for the first time, into direct contact with Fleischmann and Pons.

One of the most important issues that surfaced in Los Angeles was the status of Utah's helium measurements. Fleischmann and Pons were expected to describe their latest data on fusion-produced helium in their electrolysis cells. Initially Pons would only say that they were not ready to talk about it, leaving the audience dissatisfied and disbelieving. Earlier in the Walling and Simon's theoretical paper (see Chapter III) it was stated, based on information from Pons, that an amount of ^{4}He in agreement with the reported excess heat data had been produced in the Utah electro-

71

chemical cells. The helium was assayed by mass spectroscopic measurements of the escaping gases. In discussions at the Los Angeles meeting Fleischmann and Pons "backed away from earlier claims of neutron and ^{4}He detection, citing shortcomings in instrumentation" (C&EN, May 15, 1989). However, their admission that their earlier helium results were incorrect did not settle the helium question, because these analyses involved the cell gases and were based on the false assumption that the helium, if formed by fusion, would come out of the palladium electrodes. At Los Angeles, scientists from the Sandia National Laboratories and the Massachusetts Institute of Technology were among those who requested that the University of Utah release samples of active palladium cathodes to others for helium analyses. Pons replied as follows to this request: "[We] are taking steps to have these samples analyzed" (Wall Street Journal, May 10, 1989). Incredibly, the same request had been made previously of Pons at the April 12 ACS meeting in Dallas and now his answer was much less definitive than the one he had given four weeks earlier in Dallas when he had said that samples of their palladium rods had already been sent for analysis. Informed observers could only wonder what was going on. Any major laboratory would have been able to test for helium in a few days and resolve conclusively whether the helium concentration was commensurate with the excess heat.

Fleischmann and Pons' handling of the analyses of their claimed fusion-produced helium was a fiasco. Offers to do the analyses from reputable laboratories were refused, possibly due to patent agreements or other more subtle reasons. It was reported by University of Utah Vice President James Brophy on May 12 that the British company, Johnson-Matthey, had taken some of the palladium rods to analyze them for the presence of helium and other fusion products (Salt Lake Tribune, May 13, 1989). It is known that Johnson-Matthey had entered into an agreement with Fleischmann and Pons to supply palladium rods, on the condition that the rods be returned to them for analysis after use. No information on their analytical results on helium or any other fusion product has been forthcoming up to now. The stalling of the helium analysis was a blatant example of techniques used by the University of Utah to keep information away from the scientific and public communities. Some insight into the problem is revealed by University of Utah Vice President Brophy's statement, "I can't imagine any reasonable person turning them (the palladium electrodes) over to a competitor . . . that would drive our lawyers up a wall" (Salt Lake Tribune, May 13, 1989). This, however, is not the end of the helium fiasco, which will be continued in Chapter VIII.

The meeting of the electrochemists in Los Angeles did little to curb the swelling tide of disbelievers. The lack of any new information from Fleischmann and Pons and the secrecy surrounding the analysis of their palladium electrodes for helium caused the electrochemists in Los Angeles to become impatient and skeptical. The *Los Angeles Times* in reporting on the meeting summed up the situation with its May 9 and 10 headlines "Scientists Fail to Offer New Evidence of Fusion" and "What Started as a Triumphant Saga Fizzles into Confusion and Carping." The *Washington Post* (May 9) was still more critical and stated that Pons and Fleischmann were "facing a now almost overwhelming consensus among scientists that . . . cold fusion claims were false." Since five million dollars of state funds were being debated in Utah, University of Utah officials were sensitive to the unfavorable press coming from both the Baltimore APS meeting and the Los Angeles Electrochemical Society meeting. University of Utah spokesman Brophy tried to downplay some of this negative publicity by stating that they were fighting an "Eastern Press" that was unwilling to acknowledge that the Los Angeles meeting included several people who had confirmed the experiment (*Salt Lake Tribune*, May 11, 1989). University of Utah President Peterson added, "The University's loudest critics have come from large, well-financed schools that have sizeable stakes in the more conventional, and more expensive, hot-fusion projects" (*Salt Lake Tribune*, May 12, 1989). Shifting the blame to the 'Eastern Establishment' was an often-used technique in the cold fusion saga that played well in Utah. Another technique was to use the Japanese as a threat. In the above article, Brophy is quoted as saying "We do know the Japanese are spending time on it right now. I hate to keep using the Japanese as bogeymen, but there ought to be a national response." Recall that the University of Utah in its appeal for federal funds at the Congressional Hearing also used the Japanese to build its case (Chapter V). Available sources indicated, however, that there was at this time no special national effort on cold fusion in Japan [*Nature*, **339** 167 (1989)].

Some of the secrecy surrounding the helium analysis and other aspects of the Fleischmann and Pons research was probably mandated by the University's legal staff in order to protect its ties with business. Fleischmann had indicated that they were working on a sensitive subject and the intimate details of their work could not be released. At the Los Angeles meeting, for example, participants were prohibited from recording the sessions or photographing the researchers' graphs. The University's public relations office had put out information about businesses interested in its nuclear fusion research. The *Salt Lake Tribune* (May 13, 1989) reported that inquiries from 120 firms had been forwarded to the University's office of technology transfer, including thirty-three from large, multinational Fortune 500-type companies. Some fifty companies, including Westinghouse, had signed confidential disclosure agreements. Such agreements entitled companies to examine the University's patent applications in order to see if they wanted to pursue joint development with the school. Since such agreements didn't involve financial commitments, they cannot be taken very seriously. The University of Utah, however, has used such publicity in seeking federal funding. As described in Chapter V, University officials argued in Washington that funding through normal scientific channels was too slow to maintain an advantage over foreign competitors, especially the Japanese. Although this strategy was ultimately successful at the state level, it did not succeed at the national level.

The third conference which took place in Santa Fe during the second month (May 23–25, 1989) was devoted entirely to the subject of cold fusion and was hosted by the Los Alamos National Laboratory and the Department of Energy. Although some of the preliminary publicity had given top billing to Fleischmann and Pons, they had both decided not to attend the Santa Fe meeting. By this time both Fleischmann and Pons were becoming reclusive, avoiding both important scientific meetings and reporters whom they viewed as fundamentally hostile. What role the recommendations of the Utah lawyers and the negative publicity surrounding the dearth of new results at the Los Angeles meeting had on Fleischmann and Pons' decision is open to speculation, but no doubt was of major importance. Most of the other major participants in cold fusion were present at the Santa Fe Workshop, including a number of people from outside the United States. Among the more than four hundred attendees at the Workshop were scientists from national laboratories, universities and industry, as well as entrepreneurs hoping to exploit cold fusion. The proceedings of the Workshop were disseminated widely via live broadcast by satellite television. There were five half-day plenary sessions where some thirty-five papers were presented orally. In addition, there were evening sessions on May 23 and 24. The organizers had asked J. Bockris to introduce the discussion on the first evening with a selected number of speakers to present short contributions. He had proceeded to schedule only speakers who were claiming positive results. Bockris and other advocates of cold fusion took the position that only positive results should be presented at these sessions. Only very strong protests by the Workshop participants persuaded the organizers to include one contribution with null results. The oral presentations were supplemented with approximately eighty contributions selected for the poster sessions, a large fraction of which reported negative results.

The Santa Fe meeting brought together for the first time a wide variety of people with different interests in cold fusion, including electrochemists, chemists, nuclear and atomic physicists, materials scientists, representatives of petroleum and power companies, venture capitalists and government officials. This meeting also presented an outstanding opportunity for the Department of Energy Panel members to listen to claims and counterclaims and rub shoulders with the proponents and detractors of cold fusion. In the spirit of promoting cooperation between the chemistry and physics communities, a representative from each discipline was chosen as Workshop co-chairs. These were Norman Hackermann, a chemist, and J. Robert Schrieffer, a physicist.

The Santa Fe Workshop will go down in history as a most unusual scientific conference where the experimental claims were contradictory and chaotic. One group reported the production of tritium, some groups reported excess heat, others reported the production of very small amounts of neutrons while still others reported completely null results with neither evidence of excess heat nor fusion products. Even for those groups reporting positive results, experiments were not always reproducible, although performed under apparently identical conditions. In addition, the reported amounts of excess heat and the intensities of fusion products were different by many orders of magnitude, strongly indicating the respective rates were inconsistent with a single process. The results reported from Texas A&M are illustrative of the chaotic state of some data presented at Santa Fe, in this case data coming from the same institution. Appleby *et al.* reported 16 to 19 watts of excess heat per cubic centimeter of palladium, while Wolf *et al.* reported both a neutron intensity of four-tenths disintegration per second (equivalent to approximately 5×10^{-13} watts per cubic centimeter of palladium) as well as a tritium yield of some 10^{14} atoms in a Bockris cell run for about ten hours (equivalent to approximately 10^{-3} watts per cubic centimeter of palladium). Hence, from the above claimed measurements of excess heat, tritium and neutrons one calculates relative reaction rates in the ratios of $1/10^{-4}/10^{-14}$, respectively. This very large disparity between the excess heat and the neutron production has been a major unsolved problem dating from the time of the first press conference. Now at Santa Fe, this second discrepancy arose, namely, the discrepancy between the yields of tritium and neutrons, which are expected to be approximately equal (see reactions (1a) and (1b) on p. 6). Wolf *et al.* reported tritium results from one cell which had been analyzed also for neutrons. The reaction rate for tritium (reaction 1b) in this cell was also many orders of magnitude larger than that for neutrons (reaction 1a).

Assuming that the neutrons and tritium were indeed produced by the nuclear reactions (1a) and (1b), one must consider what is known about these reactions from so-called hot fusion and what is predicted as the deuteron temperature approaches thermal temperatures at which cold fusion is claimed to occur. These subjects were explored in a theoretical paper by G.M. Hale *et al.* In summary, Hale showed that reactions (1a) and (1b) occur with approximately equal intensities and that their cross-section ratio changes only by a few percent when this ratio is extrapolated from the energy region where it has been measured, down toward zero energy. These authors concluded that for the D+D reaction there "was no evidence for a strong enhancement at lower energies of the proton branch due to something analogous to the Oppenheimer-Philips effect." A number of proponents of cold fusion had appealed incorrectly to the Oppenheimer-Philips process to 'explain' their tritium. At very low energies below the Coulomb barrier, two deuterons at close proximity tend to orient themselves so that their constituent neutrons approach each other keeping their protons apart. This nuclear polarization enhances slightly the rate of the (d,p) reaction relative to the rate of the (d,n) reaction and is a possible mechanism for increasing slightly the tritium production relative to the neutron production. This effect for deuteron-induced reactions has been known for many years. The magnitude of this effect, however, is extremely small for the reaction of deuterium nuclei. Therefore, the large ratio of the tritium to neutron intensity reported at Santa Fe could not be the result of the D+D reaction. Such a reported large enhancement in this ratio would violate all that is known about nuclear physics.

The relatively large amounts of tritium, although orders of magnitude less than the excess heat, presented a challenge that was not resolved during the meeting at Santa Fe. Attendees were willing to believe that tritium was being measured; however, there was sharp disagreement about its origin. Several people suggested that the tritium might be contamination, and Richard Garwin suggested that the cathodes be analyzed for tritium impurity. It took more than a year for this simple check to be made as will be described in Chapter VIII. The result was that the palladium cathodes from a particular supplier did indeed contain tritium contamination. This discovery cast considerable doubt on whether any of the reported tritium in electrolytic cells was coming from nuclear fusion. More information about the claimed production of tritium will be given in Chapter VIII.

Little new positive data on excess heat was presented at Santa Fe. One of the few groups reporting new data was the Appleby group of Texas A&M University. In the opening paper of the conference, this group reported anomalous heat effects amounting to 16 to 19 watts per cubic centimeter of Pd. This was a somewhat surprising result in that it was even larger than most of the published results of Fleischmann and Pons [J. Electroanal. Chem. 261 301 (1989)], which ranged for six cells from 0.1 to 1.6 watts per cubic centimeter of Pd. Three additional cells of Fleischmann and Pons gave larger values; however, these quantities of excess heat had not been directly measured but resulted from a rescaling. Melendres and Greenwood of the Argonne National Laboratory criticized the Fleischmann-Pons' rescaling procedure in a poster by stating that these large values were due to "projections of highly questionable validity." Some words of caution must be sounded also about the large excess heats of Appleby et al. Their values of 16 to 19 watts per cubic centimeter of Pd were not measured directly but rather extrapolated from a cathode with a very small volume of 0.002 cubic centimeters, requiring the measured excess heat to be multiplied by a factor of 500! Hence, what appears to be a very large total excess heat, can disappear completely by assuming either a ten percent error in the power measurements or some 25% recombination of the D_2 and O_2 gases.

If the published data of Fleischmann and Pons are tabulated in terms of the percentage of excess power (i.e. excess power/input power), then for current densities of 8, 64 and 512 milliamperes per centimeter squared, respectively, the excess power values are 30, 28 and 21%. These are rather modest values compared to the 300% excess power announced at the initial press conference, and restated in many subsequent media reports, where it was reported that four watts of output power resulted from one watt input.[12] The Huggins and Appleby groups both reported only modest amounts of excess power, when calculated in terms of the ratio [(output power) − (input power)] / (input power), of approximately 12% and at no time had either group claimed that it was due to nuclear fusion.

Bockris et al. presented a paper in which several mechanisms were examined as possible chemical explanations of the excess power first reported by Fleischmann and Pons. Bockris et al. concluded that chemical explanations could account for no more that 2.1 watts per cubic centimeter of palladium whereas, he assumed, a successful Fleischmann-Pons' experiment produced approximately ten watts of excess power per cubic centimeter of palladium. On this basis, he concluded that the excess power was too large to have a chemical explanation. One might question Bockris' conclusion which is not based on Fleischmann and Pons' published data quoted in an earlier paragraph (0.1 to 1.6 watts per cubic centimeter) but on an *assumed* excess power of ten watts per cubic centimeter. Furthermore, the Bockris' logic, like that of Fleischmann and Pons, was flawed. Rather than use the observed nuclear products as the evidence for nuclear reactions, these proponents of cold fusion instead argued for a nuclear process on the basis of the elimination of other possible heat-producing processes.

There were a number of contributions in Santa Fe where the authors concluded that the electrolysis of D_2O with a palladium cathode did not produce any excess heat. The most important papers on calorimetry at Santa Fe were presented by two Canadian groups who used closed calorimetric cells as opposed to the open cells of Fleischmann and Pons. M.E. Hayden et al., of the University of British Columbia in Vancouver, developed a calorimeter suitable for investigating cold fusion in electrolytic cells that achieved a claimed 0.3% accuracy in absolute energy balance. Their cells used self-sustained catalytic recombination of the evolved gases and operated in a completely closed fashion. When using closed cells, one eliminates the need to make the major assumption associated with open cells that the D_2 and O_2 gases are vented without reaction. Hayden et al.

---

[12] See footnote 11.

showed no excess heat generation for input powers in the 4 to 18 watt range. Similar null results with closed-cell calorimetry were reported by J. Paquette et al. of the Chalk River Nuclear Laboratories in Chalk River, Ontario. They ran electrochemical cells under a variety of conditions and the power in and power out agreed to within a reported 2%. The question on everyone's mind at this time was why the proponents of cold fusion continued to use open-cell calorimetry when closed-cell calorimetry was obviously more accurate and more reliable.

The results of two additional high quality calorimetric measurements with electrochemical cells, both giving null results, are also included here. Redey et al. of the Argonne National Laboratory used a constant heat-loss rate calorimeter with current densities up to 500 milliamperes per square centimeter. They observed no excess heat in any of their many cells. Randolph et al. of the Westinghouse Savannah River Company also reported no excess heat in their poster paper that described precise calorimetry with very good accuracy at a power level of 10 watts.

Most proponents of cold fusion reporting excess heat from their electrolysis experiments were claiming that one of the main characteristics of cold fusion was its irreproducibility and sporadic occurrence (even though some proponents claimed success rates of up to 90%). Usually in science, irreproducibility of results is a negative signal, warning that something is fundamentally wrong with the experiment. The proponents' technique of attempting to turn the tables by making irreproducibility a positive feature of the phenomena has been indeed shocking. Another bizarre belief of the cold fusion proponents is that it is perfectly acceptable to decouple the magnitudes of the excess heat and the fusion products. Even though they claim the excess heat is due to nuclear fusion, the proponents gave no serious consideration to the fact that the fusion product intensity was orders of magnitude less than the excess heat. Fleischmann and Pons solved this glaring inconsistency very simply by stating that most of their claimed excess heat was due to an unknown nuclear process! Bockris, and others, also accepted the large discrepancy in the reported ratios of the yield of tritium to neutrons as a fundamental characteristic of cold fusion, incorrectly justifying these large ratios on the basis of the Oppenheimer-Philips process. Most of the proponents of cold fusion who reported excess heat at Santa Fe and insisted that its magnitude was so large that it must be due to nuclear fusion, unaccountably ignored well-known results in nuclear physics. They should have considered this source of information also in the evaluation and interpretation of their results.

One of the claims that had circulated from the time of the first press conference announcing cold fusion was that it is possible to electrochemically compress deuterons in palladium host lattices to such a degree that nuclear fusion would occur (see earlier discussions about the postulated large pressures of some 10^{27} atmospheres). Therefore, it was timely that several papers at Santa Fe discussed the question of the D–D distances of deuterium, when loaded to various D/Pd ratios in a palladium lattice. Peter Richards of the Sandia National Laboratories in Albuquerque presented a molecular dynamics calculation of palladium deuteride in order to answer the basic question "How close can deuterons get" when loaded in a palladium lattice. He concluded that the D–D separation distances in palladium could approach the distance of 0.074 nanometers, the bond distance in D_2 gas molecules, but never get any closer. Others, including F. Besenbacher of the University of Aarhus and J. Mintmire et al. of the Naval Research Laboratory, reached similar conclusions. On this subject Peter Hagelstein of MIT asked the question on everyone's mind: "Does anyone have a theory that will allow two deuterons to come close enough together that they have a good chance of fusing?" No one in the audience gave an answer!

A major fraction of the many experimental papers at the Santa Fe Workshop described searches for the different fusion products, including neutrons, protons, tritium, ^3He, ^4He, x-rays and gamma rays. The authors of most of these contributions reported null results in the form of upper limits on fusion products; these upper limits were orders of magnitude below the published claims of Fleischmann and Pons. Although this overwhelming evidence was interpreted by most Workshop members as evidence that the Fleischmann-Pons phenomenon was not due to nuclear fusion, a small group desiring cold fusion to be a reality continued to raise the rhetorical question "but suppose it is true?" The lure of an unlimited supply of cheap, clean and safe energy was so great that some simply refused to accept the consequences of the above stringent upper limits that were placed on the fusion product data. This group conjectured that perhaps by improving the electrochemical technique, some useful amount of fusion energy would result. At the time of the Santa Fe Workshop, the most definitive result on fusion products was the very small upper limit that had been placed on the neutron intensity (i.e. at the Jones level or below) produced during the electrolysis of D_2O in various electrolytes with palladium and titanium cathodes. This limit showed conclusively that nuclear fusion was not the source of the claimed excess heat.

Most of the discussion about fusion products centered on the very small, but positive, neutron intensity reported by Jones and his collaborators. Although the proponents of cold fusion used the Jones-like evidence to support their claim that the excess energy was due to cold fusion, the two phenomena differ in reaction rates, if fusion is assumed, by over 13 orders of magnitude (10^{13})! The failure of many proponents of cold fusion to recognize this fundamental difference between the BYU and University of Utah claims has created much confusion. Aside from the BYU group, others who reported some evidence of trace intensities of random neutrons at Santa Fe were the groups of A. Bertin of the University of Bologna (in collaboration with the BYU group) and K. Wolf of Texas A&M. Jones reported at Santa Fe a reaction rate some 6 times smaller than his published value of 10^{-23} fusions per deuteron pair per second. His earlier published rate was based on one run giving a maximum fusion rate of 0.4 per second, whereas his rate given at Santa Fe was based on several runs that gave an average rate of 0.06 fusions per second. The Bertin et al. experiment was very similar to the earlier BYU experiments utilizing titanium electrodes and the "mother-earth soup" type electrolyte;[13] however, this work was done in the Italian Gran Sasso Massif Laboratory which is under the 1400-meter Mount Aquila. Since the Bertin experiment was performed deep underground, it was surprising that the neutron background was as large as reported. The significance of the Bertin et al. and Wolf et al. experiments was that their neutron intensities were of the same order of magnitude as the Jones experiment, although slightly larger, and still near the instrument limit of detectability in each laboratory.

All three of the above positive neutron results were at intensities below the upper limits set by the vast majority of fusion product searches that reported null results at Santa Fe. However, two groups at Santa Fe reported results where the upper limits on the neutron intensity in cold fusion experiments were well below the positive result of Jones. One of these groups was the Yale-Brookhaven National Laboratory collaboration headed by M. Gai. He presented their results in Santa Fe as he had done at the Baltimore meeting. This group pushed the upper limit on the neutron intensity from electrolytically-induced fusion more than an order of magnitude (10^1) below the Jones' value, utilizing a sensitive neutron detection system with a very low background. Yale's Gai raised serious questions about all the positive results reported at Santa Fe. Measurement of low neutron levels in the presence of relatively high neutron backgrounds is a challenging experimental problem. Hence, in all cold fusion neutron experiments it is critical to reduce the background by various means to as low a level as possible. Although the Bertin et al. experiment was performed deep underground, as mentioned previously, the neutron background (probably due to gamma ray contamination) was more than an order of magnitude larger than that of the Yale-BNL collaboration. If one defines the merit of a low-level neutron experiment by the ratio of the neutron counting efficiency to the neutron background, the Yale-BNL experiment is superior to the Bertin et al. experiment. During the discussion at Santa Fe, Jones raised

---

[13] The BYU group, with their interest in low-level cold fusion in nature, used an electrolyte prescription that contained salts typical of volcanic hot springs and included electrode-metal ions. The exact complex prescription is given in their first publication [Nature 338 737 (1989)].

the question about whether the Yale-BNL collaboration was doing the right electrolysis experiment. He stated that the group was not using electrodes similar to his that were made of sintered titanium and palladium powders. Such allegations of failure to perform the electrochemistry correctly have often been leveled against those who do not find evidence for excess heat or fusion products. To settle this issue between Jones and Gai, Gai publicly invited Jones to repeat the BYU experiment at Yale, and Jones accepted the offer. This collaborative experiment between groups that had reported positive and negative results was arranged at the Workshop and was exactly the type of collaboration that was subsequently recommended by our DOE/ERAB panel. Results of this collaboration will be described in Chapter VIII.

A truly remarkable experiment that had a merit factor which was orders of magnitude better than previous experiments was described by Yves DeClais of Annecy-le-Vieux, France. He was the spokesman for a collaboration of several French groups denoted as the "Bugey collaboration." Their experiments were performed in the France-Italy Frejus tunnel where the neutron background was two counts in five days! No neutrons above this very low background were observed, and the authors reported an upper limit to the neutron intensity that was yet another order of magnitude below the Gai et al. result! Declais, anticipating criticism of his electrolysis procedures, made a standing offer at the Santa Fe meeting. He invited anyone with a serious interest in searching for cold fusion neutrons to bring their electrolysis cells to the Frejus tunnel and make use of their sophisticated neutron detector in an extremely low-background environment. At this writing, well over a year after the Santa Fe Workshop, no one to my knowledge has accepted DeClais' most generous offer. (Also not as of July 1, 1991.)

Another intriguing question that has been raised about the reported low-level neutron intensities, which have been claimed to be associated with the loading of deuterium into metal lattices, is whether or not some, or all, of these claimed particles are emitted in bursts. Several claims and counterclaims on bursts were described in Santa Fe. Two Italian groups and one American group reported neutron bursts. These groups were Scaramuzzi et al. from Frascati, Gozzi et al. from Rome and Menlove et al. from Los Alamos (with collaborators from BYU). Each of the three groups reported on a new class of experiments which were substantially different from the original Fleischmann-Pons' experiments that depended on the electrolytic loading of deuterium into metal cathodes. The new experiments utilized a dynamic mode of operation in a highly non-equilibrium state and loaded the D_2 gas into metal shavings of palladium or titanium by pressure and temperature excursions (i.e. by cycling the temperature over hundreds of degrees in a short period of time). The Menlove et al. intensity of burst neutrons was considerably smaller than that reported by Scaramuzzi et al. who were using boron trifluoride (BF_3) neutron counters. These counters are known to be notoriously sensitive to temperature and humidity changes as well as to noise. Hence, Gai of Yale made a strong plea at Santa Fe for experimenters not to use BF_3 counters in cold fusion experiments.

Several groups searching for bursts of particles from non-equilibrium "Frascati-type" experiments reported negative results at Santa Fe. Barwick et al. from the University of California, Berkeley, searched for ^{3}He produced by the fusion of two deuterons with a highly efficient plastic detector. They obtained an upper limit that was more than 2 orders of magnitude (10^2) lower than that of Scaramuzzi et al. Negative results on neutron bursts were obtained and reported at Santa Fe by Hill et al. (Iowa State), Agrait et al. (University of Madrid), Schirber et al. (Sandia National Laboratories) and McCracken et al. (Chalk River National Laboratories).

There were discussions also at Santa Fe about the possible origin of neutrons at very low intensities from processes other than cold fusion. Cosmic ray induced muon-catalyzed fusion of deuterium had definitely been ruled out by the experiments of Nagamine et al. (Japan collaboration) who used accelerator produced muons to bombard cold fusion cells and found that the muons were preferentially captured by palladium. A process known as "fracto-fusion" was invoked in Santa Fe (e.g. by F.J. Mayer et al.) as a possible source of very small bursts of neutrons, claimed to arise from a fracturing process producing cracks in the palladium and titanium metals. Such a process was first postulated in the Russian literature. For example, Boris Deryagin, well known in the United States for his erroneous work on polywater, and his colleagues have published extensively on the fracture of solids where the surfaces which are formed in the course of the fracture are electrically charged. Acceleration of deuterons across these cracks with localized charge imbalance was postulated to cause fusion. Neutrons from such a process are not due to cold fusion but claimed to be due to microscopic pockets of hot fusion. This subject will be revisited in Chapter VIII.

Another of the many variables in cold fusion experiments is the degree of loading of deuterium into the palladium and titanium cathodes. Proponents of cold fusion at the Santa Fe meeting disagreed about the importance of deuterium loading and some didn't even report the D/Pd atom ratio in their experiments. Electrolysis experiments lead to deuterium loadings where the D/Pd ratios can be as large as 0.7 to 0.8. Much larger D/Pd ratios can be obtained when loading by ion implantation. For example Myers et al. of the Sandia National Laboratories, using ion implantation, employed D/Pd ratios of 1.3 in their cold fusion experiments. Interestingly, Fleischmann and Pons themselves have never disclosed their D/Pd values, again adding to the problems in both replicating their experiments and in evaluating their results.

There were approximately thirty theoretical contributions at the Santa Fe Workshop. A number of these papers assumed cold fusion was true and then proceeded to 'explain' how the phenomenon might occur, as had a number of theoretical papers at the Baltimore meeting. Such highly speculative theories in cold fusion have often had detrimental side effects by giving hard-core fusioneers unwarranted optimism. In addition, these theories often have had short lifetimes and, in some cases, the same authors have later even come to directly opposite conclusions. Rafelski exemplified such a turnabout in Santa Fe. He stated in Baltimore that the fusion rates observed by the BYU group could be accounted for by standard nuclear physics data and WKB penetration integrals. In Santa Fe he came to the opposite conclusion, namely, "screening is unable to explain, in terms of conventional physics, the observed fusion rates, and some resonance or off-equilibrium behavior will need to be invoked." Hence, Rafelski finally concluded that the theoretical fusion rates at room temperature were much smaller than the Jones *et al.* result (a conclusion most theorists supported) and only some unproven non-equilibrium process, e.g. fracto-fusion, could account for the claimed low-level neutron intensities.

During the two and one-half day workshop in Santa Fe, our DOE/ERAB Panel had the opportunity to participate in a crash course covering all the ramifications of the cold fusion phenomena, experiencing first hand the exchange of claims and counterclaims between many of the active workers in cold fusion. My impressions of the status of cold fusion at the end of the workshop, two months after the initial press release, are best summarized by responding to the following hypothetical question. During this two-month period, had any carefully executed experiment, in which the calorimetry and fusion products were measured along with the necessary checks and controls, showed a positive result with statistics that were significant? If one required that the excess heat be accompanied by a commensurate amount of fusion products, the answer was categorically No! By this test, which many felt was a reasonable one, the large inconsistency between the reported intensities of the excess heat and fusion products was proof that a single nuclear process could not account for the origin of the two very different claims.

If one de-coupled the claims of excess heat and fusion products, and asked the above question for each claim separately, the answers at the time of the Santa Fe Workshop needed some qualifications. First, consider the topic of excess heat. The best and most careful calorimetric experiments with the better controls using closed cell calorimetry had reported no excess heat. These and other negative experiments, however, had not completely settled the question of excess heat because a few groups continued to report excess heat with open cell calorimetry, even though these positive experiments were plagued by uncertainty and irreproducibility. The magnitude of the excess heat claimed by these positive experiments was *markedly reduced* from the initial claims of Fleischmann and Pons made at the March 23, 1989 press conference. One thing came through loud and clear at Santa Fe. There was no evidence presented to support the Fleischmann-Pons claim that the excess heat was due to a nuclear process. Fleischmann and Pons in their published paper had stated that the excess heat was due to an unknown nuclear process! Such speculation is neither scientific nor warranted in the mature discipline of nuclear physics, which is over fifty years old.

Secondly, what was the status of the fusion product claims after two months? Again, the two best neutron experiments with the highest figure of merit (ratio of efficiency to background) had found nothing, and had established an upper limit for neutron production of more than an order of magnitude below the reported value of Jones *et al*. The figure of merit of the DeClais *et al*. experiment was 500 times superior to the Jones et al. experiment! Hence, the only way to discount these experiments was to postulate that the electrochemistry was faulty. Of the approximately five claims for production of either random or burst neutrons at the time of the Santa Fe meeting, all reported that the neutron intensities were very small, many orders of magnitude below the heat claims. Hence, based on the neutron yields it was clear that D_2O electrolysis was not a useful nuclear energy source. No convincing evidence had been reported either for any charged particles, gamma rays or x-rays. Although one group (Texas A&M) reported tritium at Santa Fe, contamination had not been ruled out and, furthermore, Ziegler *et al*. (IBM) and Barwick *et al*. (Berkeley) employing charged-particle detectors reported an upper limit for tritium production below the Jones *et al*. neutron intensity. Insofar as no one at Santa Fe was able to come up with a mechanism to allow two deuterons to come close enough together at room temperature to give a fusion probability even of the order of the reported very small neutron yield, discussion turned to mechanisms for producing neutrons other than cold fusion, such as fracto-fusion.

The above paragraphs summarize the state of the cold fusion phenomena when our DOE/ERAB Panel began its work in earnest. Although Panel members had been investigating cold fusion claims already for over a month at the close of the Workshop, efforts had to be intensified since the preliminary report was due at DOE headquarters in six weeks. At a meeting in the LaFonda Hotel during the evening of May 24, with most panel members in attendance, the group decided to make visits to the laboratories of the principal proponents of cold fusion as soon as possible. The first visit was to be to the University of Utah. Prior to the opening of the scientific session on May 25, I contacted Dr. James J. Brophy, Vice-President for Research at the University of Utah who was attending the Santa Fe meeting, in order to arrange the panel's visit. He was cooperative and proceeded to arrange a meeting by telephone with Fleischmann and Pons. We preferred to have the meeting early the following week prior to Fleischmann's departure for Southampton. Later that morning I received a telephone call from Pons stating that any visit of the panel to his laboratory was contingent on either dismissing several panel members that he felt were hostile or adding a number of additional individuals who he felt were supporters of cold fusion. I assured him that the panel was an excellent one representing all the pertinent scientific disciplines and comprising members who were capable of evaluating the scientific evidence without personal prejudices. After informing Pons that I considered the panel membership to be fixed, I reluctantly agreed to inform the panel of his objections. A quick meeting of the panel was held in a back room of the Sweeney Convention Center before leaving Santa Fe, and the members unanimously agreed with me that the panel membership would not be changed. Hence, the tentatively arranged meeting to the University of Utah was cancelled.

VII

Publication of the Panel's Report

Following the original announcement of the Fleischmann-Pons calo-rimetric results, many scientists performed experiments of the same general type in an attempt to verify the surprising reports of excess heat. Most of these experiments failed to reveal any excess heat within the errors of the measurements. However, Fleischmann, Pons and Milton E. Wadsworth (Professor of Metallurgy and Dean of the College of Mines and Earth Sciences) at the University of Utah, Bockris, Appleby and their colleagues at Texas A&M, Huggins at Stanford and a few others continued to main-tain that they were seeing excess heat. In the six-week period before its interim report was due, our DOE/ERAB panel chose to visit research groups at six laboratories to observe first hand the methods and equipment being used in making calorimetric and fusion-product measurements. Included in these visits were groups at the three above universities, well-known for their reports of excess heat; the Jones group of Brigham Young University who claimed neutron emission at a level thirteen orders of magnitude less than the University of Utah claim of excess heat; the McKubre group of Stanford Research Institute (SRI) with an active calorimetric program (visit took place at the Electric Power Research Institute); and the Lewis-Barnes group of chemists and physicists at the California Institute of Technology which had concluded, on the basis of an extended series of calorimetric and counting experiments, that neither excess heat nor nuclear particles were produced from electrochemically charged palladium cells.

Immediately after returning from the Workshop on Cold Fusion in Santa Fe, I initiated new discussions with Brophy that resulted in scheduling a visit by several panel members to the University of Utah on June 2, 1989. Our panel was especially interested in seeing the Fleischmann-Pons cells in operation and in learning first hand about the cell calibration which was so important in the calculation of their output power. It turned out, however, that Pons was unable to locate the desired critical cell calibration data during our panel's visit! Since the outcome of whether or not excess power was being produced depended directly on the cell calibration, the panel's visit to the University of Utah was unsuccessful in this regard. Pons did, however, promise to send their calibration data to us in the next few days well before we had to write our interim report. This promise to send the crucial calibration data to our panel never materialized. During its visit to the University of Utah, our panel also stopped at the laboratory of Milton E. Wadsworth who had also reported excess heat. At Texas A&M, our panel visited four separate groups involved with cold fusion research: those of Appleby, Bockris, Martin and Wolf.

From the point of view of seeing a working fusion cell, our panel's visits to the different sites were very disappointing. In none of our meetings with the various groups claiming positive results did we see an operating cell that was producing excess heat at the time of our visit! The panel's site visits did identify, however, a number of uncertainties and problems associated with some of the experiments, including questionable calibration procedures and doubtful accuracy of data acquisition. These problems were especially critical because in many cases, even when experiments were run under identical conditions, the positive results were not reproducible. The proponents of cold fusion offered many reasons why most experimenters who had performed calorimetry reported no excess heat. They maintained that the failures were due to an insufficient time of electrolysis, too small current densities, deficiencies in the materials, impurities, etc. When one examined the results of those reporting positive results, however, one found that a wide array of cell parameters were used. One proponent reported the necessity to exceed a threshold current density of 100 milliamperes per square centimeter in order to generate excess heat, while others claimed excess heat from comparable cells with current densities of 8 and 64 milliamperes per square centimeter. Some proponents stated that very long electrolysis times were required for initiation of excess heat production, while others with similar cell configurations claimed success after a few hours of electrolysis. One was forced to conclude at this time that no one, including the proponents, had clearly identified the differences in the cells, materials or operating conditions that could account for the highly irreproducible results from cold fusion experiments. This irreproducibility and sporadic character of cold fusion, admitted to by even the strongest proponents of cold fusion, has served as one of the grounds for the skeptics to reject cold fusion as valid science.

During its deliberations the panel had available to it reams of published and unpublished reports and communications from laboratories both inside and outside the United States. Although almost all of the important work of the panel, including the writing of preliminary drafts of the reports, was performed by subpanels working diligently over the six-month period, the tone and final content of the interim and final reports were established during four meetings of the entire panel. These intense meetings were open to the public and allowed the press to witness firsthand the sometimes differing views of panel members. Carrying out our many deliberations and report writing under the bright glare of television lights was not pleasant at times, but it was a situation that had been accepted by panel members as an inevitable downside to keeping our work public.

As one can imagine, the television networks, wire services and press were quick on the uptake of our panel's open deliberations. They often disseminated widely the panel's point of view even before the final wording was agreed upon by all panel members. As an example of the media's quick and accurate analysis of our panel's complex deliberations, the NBC Nightly News carried the following statement by Tom Brokaw on July 11, 1989, the very day of a panel meeting in Washington: "In the United States today, some cold water was thrown on the idea of cold fusion. A panel of twenty-two leading scientists convened by the Secretary of Energy reports it found no evidence that fusion represents a significant source of energy. And it says the government should not support any major cold fusion research projects currently under consideration."

On the next morning, July 12, the day our panel completed its interim report, *The Today Show* carried the story entitled "Panel Discounts Cold Fusion Claims."

JOHN PALMER: A panel of government-appointed scientists is reported ready to make public a report on the controversial scientific process called cold fusion.

Science correspondent Robert Bazell has been covering the investigation. He reports on the findings.

ROBERT BAZELL: Twenty-two prominent scientists assembled by the Department of Energy made up the first official group to examine the phenomenon called cold fusion. Members visited the laboratory in Utah where Dr. Stanley Pons and Dr. Martin Fleischmann claim to have created fusion in these jars, to have harnessed the reaction that fuels the sun.

Pons and Fleischmann hinted they may have found a cheap new source of energy for the world. Not so, the panel concluded in its draft report. The report said there is no convincing evidence that what is called cold fusion will ever produce significant amounts of energy.

Although many people wanted to believe Pons and Fleischmann, and some other scientists said they had repeated parts of the experiment, many scientists have predicted that cold fusion was too simple to be real.

Dr. ROBERT PARK [American Physical Society]: It's a story the American people love. They love the story about the manager of the New York Yankees who spots some hillbilly throwing rocks at squirrels, and then takes him to New York to win the World Series. But, of course, it never happened.

BAZELL: It now appears that cold fusion will be remembered as one of the memorable mistakes in the history of science.

The results of our panel's intensive study of cold fusion are well known and published in two U.S. Department of Energy Reports, the interim report DOE/S–0071 (August, 1989) and the final report DOE/S–0073 (November, 1989). The conclusions and recommendations as published in the panel's final report are reproduced below:

CONCLUSIONS

(1) Based on the examination of published reports, reprints, numerous communications to the panel and several site visits, the panel concludes that the experimental results of excess heat from calorimetric cells reported to date do not present convincing evidence that useful sources of energy will result from the phenomena attributed to cold fusion.

(2) A major fraction of experimenters making calorimetric measurements, either with open or closed cells, using Pd cathodes and D_2O, report neither excess heat nor fusion products. Others, however, report excess heat production and either no fusion products or fusion products at a level well below that implied by reported heat production. Internal inconsistencies and lack of predictability and reproducibility remain serious concerns. In no case is the yield of fusion products commensurate with the claimed excess heat. In cases where tritium is reported, no secondary or primary nuclear particles are observed, ruling out the known D + D reaction as the source of tritium. The Panel concludes that the experiments reported to date do not present convincing evidence to associate the reported anomalous heat with a nuclear process.

(3) The early claims of fusion products (neutrons) at very low levels near background from D_2O electrolysis and D_2 gas experiments have no apparent application to the production of useful energy. If confirmed, these results would be of scientific interest. Recent experiments, some employing more sophisticated counter arrangements and improved backgrounds, found no fusion products and placed upper limits on the fusion probability for these experiments, at levels well below the initial positive results. Based on these many negative results and the marginal statistical significance of reported positive results the Panel concludes that the present evidence for the discovery of a new nuclear process termed cold fusion is not persuasive.

(4) Current understanding of the very extensive literature of experimental and theoretical results for hydrogen in solids gives no support for the occurrence of cold fusion in solids. Specifically, no

theoretical or experimental evidence suggests the existence of D–D distances shorter than that in the molecule D_2 or the achievement of "confinement" pressure above relatively modest levels. The known behavior of deuterium in solids does not give any support for the supposition that the fusion probability is enhanced by the presence of the palladium, titanium, or other elements.

(5) Nuclear fusion at room temperature, of the type discussed in this report, would be contrary to all understanding gained of nuclear reactions in the last half century; it would require the invention of an entirely new nuclear process.

RECOMMENDATIONS

(1) The Panel recommends against any special funding for the investigation of phenomena attributed to cold fusion. Hence, we recommend against the establishment of special programs or research centers to develop cold fusion.

(2) The Panel is sympathetic toward modest support for carefully focused and cooperative experiments within the present funding system.

(3) The Panel recommends that the cold fusion research efforts in the area of heat production focus primarily on confirming or disproving reports of excess heat. Emphasis should be placed on calorimetry with closed systems and total gas recombination, use of alternative calorimetric methods, use of reasonably well characterized materials, exchange of materials between groups, and careful estimation of systematic and random errors. Cooperative experiments are encouraged to resolve some of the claims and counterclaims in calorimetry.

(4) A shortcoming of most experiments reporting excess heat is that they are not accompanied in the same cell by simultaneous monitoring for the production of fusion products. If the excess heat is to be attributed to fusion, such a claim should be supported by measurements of fusion products at commensurate levels.

(5) Investigations designed to check the reported observations of excess tritium in electrolytic cells are desirable.

(6) Experiments reporting fusion products (e.g., neutrons) at a very low level, if confirmed, are of scientific interest but have no apparent current application to the production of useful energy. In view of the difficulty of these experiments, collaborative efforts are encouraged to maximize the detection efficiencies and to minimize the background.

Over the six-month period that our panel had devoted to studying the volumes of published and unpublished reports and communications to it regarding cold fusion, the individual assessments of panel members had converged sufficiently to permit unanimous approval of the final conclusions and recommendations reproduced above. The last one-half day (October 31, 1989) of the panel's open meeting in Washington did, however, produce a short crisis. My Co-Chairman Norman Ramsey, had gone abroad on July 15 after our Interim Report was completed and had not participated in either the panel's further study, investigations and deliberations or the writing of the final report. Ramsey arrived at this last meeting, when editorial changes in the final report were essentially complete, and asked me (as chair of the Panel's meeting in Washington) if he could read a prepared statement. Much to my and other panel members' surprise, it was a long letter of his resignation from the panel. Ramsey's stated reasons for his resignation were related in one way or another to his lack of participation in the panel's work during the four months prior to the final report. As it developed, however, I concluded that Ramsey's main motivation for his resignation was to exert pressure on the panel to accept a preamble that would weaken the panel's conclusions. He immediately stated that he would withdraw his resignation if the panel agreed to add his already prepared preamble. It would have been damaging to the impact of our final report had Ramsey, a new Nobel Prize winner in physics, resigned as our final report was just being completed. Facing this hard choice, three hours before our panel was to disperse, the panel opted to accepted Ramsey's preamble, rather than his resignation, under the condition that his preamble could be modified slightly. Ramsey's preamble, as revised by the panel, is given below and appears in the final report prior to the conclusions and recommendations.

PREAMBLE

Ordinarily, new scientific discoveries are claimed to be consistent and reproducible; as a result, if the experiments are not complicated, the discovery can usually be confirmed or disproved in a few months. The claims of cold fusion, however, are unusual in that even the strongest proponents of cold fusion assert that the experiments, for unknown reasons, are not consistent and reproducible at the present time. However, even a single short but valid cold fusion period would be revolutionary. As a result, it is difficult convincingly to resolve all cold fusion claims since, for example, any good experiment that fails to find cold fusion can be discounted as merely not working for unknown reasons. Likewise the failure of a theory to account for cold fusion can be discounted on the grounds that the correct explanation and theory has not been provided. Consequently, with the many contradictory existing claims it is not possible at this time to state categorically that all the claims for cold fusion have been convincingly either proved or disproved. Nonetheless, on balance, the Panel has reached the following conclusions and recommendations.

In my view, the preamble is a noncommittal and evasive statement that gives credence to the claim that the experiments were not consistent and reproducible. It would, of course, be revolutionary if even one short but valid cold fusion period had occurred! However, it was just this kind of evidence that was missing in all the proponents' claims. As in all areas of science it was, of course, not possible to state categorically that all the claims for cold fusion had been convincingly either proved or disproved. Some scientists will continue to believe in an experimental result or an outdated theory even though there is solid evidence to discredit it. While the panel was willing to accept the above caveat, which weakened the report somewhat, it insisted on preserving the report in its entirety, including a forceful concluding item as given above by conclusion five (see e.g. I. Goodwin, *Physics Today*, December, 1984, p. 43). In retrospect, Ramsey's insistence on including the preamble may have been in deference to two colleagues, both Nobel Prize winning theorists in physics. Ramsey had received communications, already prior to completing our interim report, from Professors Julian Schwinger (UCLA) and Willis E. Lamb, Jr. (University of Arizona) who had invented 'explanations' of cold nuclear fusion in an atomic lattice. Both Schwinger and Lamb later published their respective 'explanations' of cold fusion and I refer to their speculations elsewhere.

The first conclusion summarized the panel's view on the claim of excess heat (or, more precisely, excess power) from electrochemically charged palladium cells. Those interested in a tabulation of groups reporting positive and negative results on excess heat as of October 31, 1989 should see Tables 2.1 and 2.2 in our panel's final report DOE/S–0073. As the report states, "In most cases calorimetric effects attributable to excess heat are small." Furthermore, "the calorimetric measurements are difficult to make and may be subject to subtle errors arising from various experimental problems." Several of the experimental problems in the evaluation of heat effects are discussed in Appendix 2.C of our final report DOE/S–0073. Calorimetry has turned out to be much more difficult than initially advertised (see, for example, Appendix I of our DOE/ERAB panel's final report). Uncertainties in such results are caused by problems with calibration, failure to run sufficient control cells, failure to properly assess errors in the primary measurements, possible electrical artifacts (one investigator had difficulties with electrical short circuits within his calorimeter, which he quickly publicized as evidence for cold fusion before fully understanding the workings of his instruments), and the assumption made about the recombination of D_2 and O_2. It is also important to recognize, as our report states,

[T]hat most of the reported excess heat measurements are actually power measurements, and the data in most experiments have not conclusively demonstrated that the total amount of energy produced (as heat and chemical energy) integrated over the whole period of cell operation exceeds the total electrical energy input.

[T]hat after assessing the reports from the different laboratories, considering the experimental difficulties and calibration problems, as well as a lack of consistency and reproducibility in observation of the excess heat phenomenon, we (the DOE panel members) do not feel that the steady production of excess heat has been convincingly demonstrated.

Pons did not agree with our panel's conclusions about the uncertainties associated with their reported calorimetric measurements, in particular, the uncertainties due to their calibration procedures and their assessment of errors in the primary measurements. He wrote to me on September 19, 1989 and stated:

[Y]ou can therefore imagine our irritation with the subsequent statements made by members of your committee regarding the accuracy of our calorimetric measurements. We established the maximum error limits for our experiments before we submitted our preliminary publication, and we did indeed give these in the paper.

Since I was unable to locate this information, I wrote back to Pons on October 16, 1989:

I presume you are referring to your Preliminary Note published in J. Electroanal. Chem. **261**, 301–308 (1989). I do not find any discussion of the different types of errors associated with open cell calorimetry in this paper. Nor do I find any information on the cell calibration and the errors associated with your particular method of calibration. Would you please inform me where this information is published?

Pons never responded. I did, however, get a letter from Martin Fleischmann written on November 9 stating that he was replying on behalf of both Pons and himself. Although I appreciated receiving his response, it did not address my above questions. Fleischmann answered in part:

It seems to be difficult for people to recognize that this paper was a preliminary publication. Such publications have strict space limitations and inevitably we could not give a detailed account of the component parts of the errors nor how we arrived at these figures. You will know that several members of your committee have wanted access to the text of our detailed paper on this subject (which is still not completed although the part you would need has been finished). As you may know we had considered meeting these requests but had to decide against this since it became clear that confidentiality could not be guaranteed.

As is so clearly illustrated above, secrecy played far too great a role in Fleischmann and Pons' dealings with the scientific community. Since all of the meetings of our panel were open to the public, it was not possible for us to guarantee confidentiality. This lack of openness was a problem also among the scientists at the National Cold Fusion Institute. Hugo Rossi, the first Interim Director of the Institute, argued for scientists at the Institute to openly exchange information with other scientists there, but Rossi was not completely successful in instituting this policy. Secrecy tends to breed careless mistakes and second-rate science and is detrimental in advancing scientific knowledge.

There have been a few changes in the makeups of the groups reporting positive and negative excess heat measurements since the panel's report. The group of Professor R.A. Oriani of the University of Minnesota was one of the groups reporting excess heat at the time of the panel's report. At that time, he received a great amount of public attention for positive results from two runs, one weakly positive (~5%) and the other yielding approximately 20% excess heat.[14] Oriani's measurements were limited, however, to calorimetry and no attempt was made to search for neutrons, tritium, gamma rays or x-rays. Since reworking its calorimeter in late 1989, Oriani's group has not had any experimental runs over the many subsequent months where an excess amount of heat was observed!

One of the more fascinating stories on cold fusion research is that of L.J. and T.F. Droege, metallurgical and electrical engineers, respectively. On learning about the Fleischmann-Pons excess heat from the *MacNiel-Lehrer News Hour*, the two brothers immediately started to build a calorimeter in T.F. Droege's basement. T.F. Droege's employer, the Fermi National Laboratory near Chicago, had forbidden him to work on cold fusion during regular working hours, but the two brothers were nonetheless sufficiently convinced of the accuracy of Fleischmann and Pons' cold fusion claims that they devoted long hours of their own time and their own finances in an effort to verify them. At one point, while working on our final report, a story circulated in the Department of Energy in Washington that cold fusion had been confirmed at the Fermi National Laboratory. While the Droeges were constructing their apparatus, they were severely critical of the many respected groups that had carried out calorimetric experiments under a wide variety of conditions and had found no excess heat. In addition, they accused high energy physicist colleagues not accepting the validity of 'anomalous heat' measurements, which comprised a large majority, of having their minds closed to new ideas. They leveled the same accusations against our panel's members. These accusations were born of conclusions T.F. Droege had made after his discussions with several high-energy physic-

---

[14] These positive results were published much later [*Fusion Technology* 18 652 (1990)] without a word of warning that in the many ensuing months all such measurements gave negative results.

ists during a 1989 summer study in Breckenridge, Colorado. Droege believed that the physicists had arrived at negative positions on cold fusion without having read the appropriate papers on cold fusion. Based on this experience with the high-energy physicists, he assumed that the DOE/ERAB panel members would behave in a similar way and remarked to me after our panel hearings: "So the conclusion of your Committee is not surprising. Collect twenty-two 'scientists' off the street and twenty of them will have minds closed to new ideas. The twenty quickly beat up the two and a report is produced like you sent." This evaluation of our panel's six months of investigation illustrates the depth of feeling of the believers in the Fleischmann-Pons phenomenon. More than once, such believers have been intoxicated and misled by a grand illusion. Laetrile for cancer and Lysenko's inheritance of acquired characteristics are recent examples. Usually the illusion centers on some much-to-be-desired wish, like the cure of cancer. Cold fusion is certainly no less grand a dream, limitless energy for mankind.

To the Droeges' credit, they completed the construction of a sensitive calorimeter and reported on their measurements during the First Annual Conference on Cold Fusion (March 28–31, 1990) in Salt Lake City. One of their goals was "to be the first to hold the land speed record for a cold-fusion-powered car." They admitted their bias and enthusiasm for cold fusion by stating "We consciously allow our excitement to show while attempting to report objectively." The Droeges reported a small anomalous excess heat at the 4% level for palladium electrodes, a result well below that reported by several other proponents of cold fusion. They added, however, "there are enough calibration runs which show too much heat and D_2O runs which show little or no heat that the whole process could be noise." Here is an example of very strong proponents of cold fusion concluding, after their measurements showed only small positive values of excess heat, that they may have observed nothing more than background! Hence, in spite of their early criticism of our panel's report, the Droeges' ran enough control experiments to conclude that they had no hard evidence for excess heat. This conclusion was in sharp contrast to that of a number of cold fusion proponents who were not nearly so careful in assessing their systematic errors. This was in part due to the temperament of the time, where calorimetric research was hurried and positive results were reported without a sufficient number of control experiments. This climate led to mistakes of various types. The fact that the reported positive results were not reproducible didn't stop the proponents from making a number of inconsistent claims.

Our panel's second conclusion summarizes the results of two categories of experimenters doing cold fusion studies. The first category, comprising most of the experimenters making calorimetric measurements, either with open or closed cells, reported neither excess heat nor fusion products. The second category of experimenters, however, reported excess heat production and either no fusion products or fusion products at a level well below that implied by their reported heat production. This discrepancy between the magnitudes of the reported heat and fusion products has been a dilemma from the time of Fleischmann and Pons' first press conference.

In the first category, among the several research groups who had done inclusive experiments searching for excess heat and fusion products, and had found neither, was the Lewis-Barnes team at Caltech, whose experiments were discussed in Chapter VI. Another such diverse group, comprised of eleven scientists working at the Harwell Laboratory in England, was headed by electrochemist D.E. Williams. This team early on had done a very comprehensive series of experiments using three different calorimetric designs and efficient nuclear particle detection on a wide range of materials and detected neither excess heat nor fusion products. During our panel's work, Williams kept panel members up to date on the results of the Harwell experiments, which have now been published in *Nature*, 342 375 (1989). Williams *et al.* suggested that undetected spurious effects such as noise from neutron counters, cosmic-ray background variations and calibration errors in simple calorimeters could have led to the Fleischmann and Pons claim of cold fusion. The negative results of the Williams group were especially meaningful because of the one-time close professional association of Williams and Fleischmann. Fleischmann had served first as Williams' Ph.D. advisor and later as a consultant to the Harwell Laboratory. In fact, Fleischmann visited Williams at Harwell on February 14, 1989 and described some of his results to Harwell scientists, some five weeks before the March 23 press conference. It was during this visit that Harwell scientists located a health physics radiation monitor for Fleischmann to take back to his laboratory. Early in March Fleischmann gave Williams instructions on cell preparation and supplied two complete cells to Williams for neutron measurements. It is my understanding that the two had limited contact, however, during most of the period that the Harwell scientists were working on cold fusion. Fleischmann did return to Harwell to present a colloquium on the University of Utah results on March 28. By this time preprints of their forthcoming article in the *Journal of Electroanalytical Chemistry* had been distributed.

On June 15, 1989, Harwell called a press conference to announce that it was terminating its research on cold fusion which it had initiated in March prior to the University of Utah press conference. The Harwell announcement stated:

> [T]he potential benefit and scientific interest in cold fusion, together
> with the Government's need for information and advice meant that the

subject had to be investigated. However, results to-date have been disappointing and we can no longer justify devoting further resources in this area . . . This work demonstrated our capacity to mount a thorough program in basic science at short notice and the ability to put together a sophisticated cocktail of scientific expertise and equipment [which] is the unique attribute of the UK Atomic Energy Authority.

It was estimated by the Harwell authorities at the press conference that their major series of experiments had cost £320,000 sterling and that the experimenters utilized £4,000,000 worth of equipment. The decision by this large multidisciplinary group, who had carried out one of the most thorough searches for cold fusion, to terminate their research work had a devastating impact on the world's opinion of cold fusion. This was especially true following the well-publicized negative conclusions reached by the Caltech experimenters.

The second category of experimenters, those claiming cold fusion based on excess heat production and either no fusion products or fusion products at a level well below that implied by their reported heat production, were faced with a striking inconsistency. These proponents of cold fusion had claimed fusion-induced excess heat but had not measured a commensurate yield of fusion products. Note that no experimentalist at all was claiming excess heat with a commensurate yield of fusion products, a result that would have been, of course, revolutionary! It is remarkable that so many of the proponents of cold fusion either ignored or paid little attention to this fact. Since the claims in support of cold fusion are faced with this large discrepancy between the measured excess heat and the fusion products, it is of interest to enumerate some of the ways the experimenters under discussion rationalized their positive results:

(1) In the cases where no fusion product measurements were made or no fusion products were observed, the excess heat was simply assumed to be due to nuclear reactions because it was too large to be chemical in origin! Several proponents of cold fusion adopted this point of view. These proponents of cold fusion reported excess heat in one or a few cells and did not give sufficient attention to the statistical assessment of their errors. Even the most ardent proponents of cold fusion admitted that the excess heat was erratic and irreproducible. It is unacceptable to compare results from many test cells against the behavior of one, or only a few, control cells as some proponents did. Statistically meaningful results required comparisons made of like numbers of cells of both types. It is very dangerous to focus on one, or a few, positive results without a thorough analysis of sufficient control experiments. Designation of the reported excess heat as nuclear in origin when no commensurate amounts of nuclear products were observed was not only illogical, but above all unscientific.

(2) In the cases where the fusion products were reported to be many orders of magnitude less than the excess heat, nearly all the excess heat was assumed to be due to an unknown nuclear process. This point of view was first stated by Fleischmann and Pons [see *Journal of Electroanalytical Chemistry*, **261**, 301 (1989)]. Their assumption that the reported excess heat was due to some unknown nuclear process puts the responsibility on them to delineate the characteristics of such a process. Nuclear physics is a relatively mature field in which the nuclear reactions of nuclei are well understood both experimentally and theoretically. During the time that has elapsed since the original Fleischmann-Pons' press conference, no new information has appeared to enlighten us about the unknown nuclear process. Therefore, at the First Annual Conference on Cold Fusion (March 28–31, 1990), I raised the issue of the unknown nuclear process with Fleischmann in a private discussion. I asked Fleischmann how it was possible for him to continue to accept a highly unlikely and unknown nuclear process as an explanation of the many orders of magnitude discrepancy between the reported excess heat and fusion products. Fleischmann's answer was immediate, indicating that he had considered such a question previously. "Confidentially", he said, "I think the palladium is fissioning"! The answer startled me. Fleischmann was, no doubt, unfamiliar with the fact that I had researched nuclear fission for decades and had coauthored a book and many research publications on the subject. Knowing that the energy threshold for palladium fission is several tens of millions of electron volts, I had to conclude that Fleischmann was either joking (although he appeared to answer in all seriousness) or exposing the gaping holes in his knowledge of nuclear physics. This abruptly ended our discussion about the Fleischmann-Pons unknown nuclear process. In relaying this story to Douglas Morrison at the Salt Lake City meeting, I learned that Fleischmann had also suggested to someone previously that the excess heat was associated with palladium fission.

(3) There were cases where variations of (2) were constructed by proponents of cold fusion to explain the large disparity between the intensities of the reported excess heat and fusion products. Some of these proponents dispensed with the nuclear reaction Coulomb barriers and branching ratios for reactions in a palladium lattice and others proposed new coherent fusion scenarios involving exotic reactions in electrochemical cells. Some went so far as to suggest that conventional nuclear physics was not applicable in solid lattices.

In all three of the above scenarios, the proponents of cold fusion attempted to rationalize the gigantic discrepancy between the reported excess heat and fusion products. In the final analysis, however, there must be an energy balance between excess heat and fusion products if a nuclear fusion process is invoked. This lack of agreement between the intensities of the excess heat and fusion products supports the panel's second conclusion that no convincing evidence exists to associate the reported anomalous heat with a nuclear fusion process.

The third conclusion in our DOE/ERAB panel report addressed the early claims of neutrons at very low levels produced from D_2O electrolysis and D_2 gas experiments. The first report of such neutrons was made by Jones and collaborators at Brigham Young University as discussed in previous chapters. Their reported neutron intensity was some 13 orders of magnitude (10^{13}) below that necessary to be commensurate with the excess heat claimed by Fleischmann and Pons. All agree, therefore, that these experiments have no apparent applications to the production of useful energy. On the other hand, if these low-level neutron results were confirmed, they would be of great scientific interest. This is because cold fusion at the Jones level still requires an enhancement of 40 orders of magnitude (10^{40}) in the theoretical D+D fusion rate, and no mechanism is yet known by which these theoretical rates could be enhanced by such a gigantic factor in order to agree with the reported observations. Although a few groups in addition to the Jones group had claimed to observe very low-level neutron intensities in either a random or burst mode, many other groups had reported no neutrons. Some of the latter groups had set upper limits well below the claims of groups reporting positive results. The DOE/ERAB panel's third conclusion, therefore, stated: "Based on these many negative results and the marginal statistical significance of reported positive results the Panel concludes that the present evidence for the discovery of a new nuclear process termed cold fusion is not persuasive." Hence, the evidence for low-level fusion products from cold fusion experiments, like those of Jones et al., is also still not established. The subject of low-level fusion products, claimed to have been produced from D_2O electrolysis and D_2 gas experiments, will be further discussed in the next chapter. However, the substance of this conclusion continues to be valid.

Our panel's fourth conclusion specifically dealt with the many speculations in the published literature and in unpublished reports that the fusion of D+D might be greatly enhanced by the presence of metals like palladium and titanium. Fleischmann and Pons, in both lectures and publications, emphasized the strange behavior of deuterons when electrochemically compressed into palladium cathodes, often stating that the fugacity of deuterium reaches astronomical values of 10^{27} (in, e.g., Fleischmann's August 5, 1989 lecture and elsewhere he stated that the effective pressure was 10^{27} atmospheres) based on simple considerations. The underlying assumption that the conditions in the metal lattice were extreme enough to alter nuclear reaction rates had apparently been a key factor in Fleischmann and

99

Pons' decision to embark on their cold fusion research. Recently, Fleischmann clearly stated that their strong belief in cold fusion dated from before the onset of their research on this subject: "We, for our part, would not have started this investigation if we had accepted the view that nuclear reactions in host lattices could not be affected by coherent processes" (*Proc. of the First Annual Conference on Cold Fusion*, p. 347, 1990).

This view that nuclear fusion could be enhanced in solid lattices was stated again by Fleischmann in another context (*ibid.* page 344). The reaction D+D→T+p (see reaction 1b on p. 6) in which two hydrogen isotopes are produced with atomic masses of 3 (tritium) and 1 (proton) was discovered in 1934 by M.L. Oliphant, P. Harteck, and Lord Rutherford (*Nature* **133**, 413, 1934). These early experiments, employing deuterated inorganic compounds as targets, were confirmed by P.I. Dee (*Proc. Roy. Soc.* A **148**, 623, 1935) using cloud chamber techniques along with kinematic momentum analysis of the fusion products. At the time of this early work the deuteron and deuterium were known as diplon and diplogen, respectively. Fleischmann stated that the early work on the D+D reaction had been largely forgotten because of the above name changes. Fleischmann said, "This neglect of the early literature is doubly unfortunate because the precise characterization of the above reaction using cloud chamber methods already showed at the time that a *significant* (my emphasis) number of tracks of the T and p species emerged at approximately 180°." Fleischmann's statement is not quite correct, however, as Dee's article (*ibid.*) shows: "*Occasionally* (my emphasis) pairs are observed in which the angle between the tracks is very near to 180° – this no doubt being the result of transmutations effected by slower diplons which have lost energy by collision in the target." Fleischmann has interpreted these very approximate kinematical coincidence data as support for cold fusion! In his words, "We believe that these reports are the first indication that there are low energy fusion channels in solid lattices." Fleischmann appears here to have succumbed to one of the pitfalls in scientific research: the manipulation of other data to fit their (Fleischmann and Pons') interpretation of their own results, an interpretation which was tailored to fit earlier preconceptions.

The idea that a metal like palladium or titanium might catalyze fusion has its origin in the ability of such metals to absorb large amounts of hydrogen (or deuterium or tritium). This special property of palladium to absorb hydrogen was known already in the nineteenth century. The two German scientists, F. Paneth and K. Peters, utilized palladium to absorb large volumes of hydrogen in their experiments in the late 1920s. Even though metals like palladium can store large amounts of deuterium, the deuterium atoms are still much too far apart for fusion to occur. As our panel's report states, the closest deuterium-deuterium distance in palladium is approximately 0.17 nanometers (nm); a distance which is large compared to the bond distance in D_2 gas molecules of 0.074 nm. Likewise as summarized in our panel report the magnitude of the deuterium pressure attained by cathodic charging of deuterium into palladium is rather modest, of the order of 10^4 atmospheres, rather than the 10^{27} atmospheres claimed by Fleischmann and Pons.

Contrary to Fleischmann's view, therefore, that D+D fusion is enhanced in palladium lattices, our panel concluded that there is no persuasive evidence for either very high pressure or very close distances of the deuterium in metal lattices. After examining a considerable amount of literature on hydrogen in solids, we stated in our fourth conclusion: "that there was no theoretical or experimental evidence that supports the existence of D–D distances shorter than that in the molecule D_2 or the achievement of confinement pressures above relatively modest levels." Hence, as discussed in our panel report, "the known behavior of deuterium in solids does not give any support for the supposition that the fusion probability is enhanced by the presence of the palladium, titanium, or other elements."

The panel's fifth conclusion that "Nuclear fusion at room temperature . . . would be contrary to all understanding gained of nuclear reactions in the last half century" summarizes our view of cold fusion in a short and forceful statement.

The panel's recommendations were of two types. First, in the area of funding, there should be no special funding for cold fusion research. This recommendation was meant to insure that no research centers or special programs to investigate so-called cold-fusion phenomena would be developed with federal funds. On the other hand, modest support was recommended for carefully focused and cooperative experiments between groups reporting positive and negative results. Our panel recommended that these funds were to be distributed within the present funding system. This phrase is well understood in American science to mean through proposals that are peer reviewed and awarded on the basis of scientific excellence. Douglas Morrison of CERN has suggested that the latter phrase should have been stated explicitly as it was in our interim report. Perhaps so, however, I doubt that any ambiguity existed among federal agencies about our meaning. Secondly, in the area of future experiments, our panel recommended collaboration between those groups reporting positive and negative results. There had been up to this time, however, no collaborative experiments between groups who reported positive and negative results on excess heat. On the other hand, a joint experiment between the groups of Jones (BYU) and Gai (Yale) was performed searching for low-level neutrons (this experiment will be discussed in the next chapter). As things are developing, it is not likely that there will be future meaningful collaboration between believers and skeptics. This is unfortunate, but inevitable, since essentially all of the skeptics have by now returned to their regular research.

101

Since the publication of our report, there has been little, if any, new hard data reported in the areas of either calorimetry or fusion products that give a better understanding of the phenomenon known as cold fusion, for which claims were made that large amounts of excess heat were produced in electrochemical cells. One of the striking features of the First Annual Conference on Cold Fusion was that the major claims of the proponents of cold fusion had remained essentially the same since the Santa Fe Conference.

The reactions to our panel's report were generally very favorable; however, not surprisingly, some of the strong supporters of cold fusion were very critical. For example, Dr. Bockris of Texas A&M wrote to me on December 26, 1989 expressing his negative evaluation of our report:

> With the recent publications describing nuclear electrochemical effects from Oak Ridge, Los Alamos, Brookhaven, and N.R.C. (sic; . . . I think he means the Naval Research Laboratory), the Fleischmann-Pons phenomena have now been repeatedly confirmed by U.S. Government Research workers. These men have added their confirmatory experiments to those obtained from April onwards at Texas A&M, Stanford, Minnesota and elsewhere. Apart from the U.S. Government labs and universities, these startling phenomena have been observed and reported on in many other countries.
>
> As to those expert and respected scientists (e.g., at Yale, MIT, Cal Tech, etc.) who denied the existence of these effects, it is now easy to understand the absence of their results: the effects, now seen by so many, do not switch on for many weeks of electrolysis. As to reproducibility, it is poor. However, for 1 mm wires, we have a 70% chance of producing tritium, if electrolysis is continued for more than 10 weeks.
>
> The situation as now seen appears markedly contrary to the impression engendered by the report of your committee. It would be fair to say that many who read it might have thought that nuclear electrochemistry was perhaps the giant mistake which it is universally pictured to be among those who do not research it.
>
> In view of the direct confirmation of these phenomena from so many sources, I put it to you that the negative ERAB report on Cold Fusion should be withdrawn. As an official U.S. Government document, it is important that it be beyond reproach (and above all, not subject to ridicule). The recent U.S. government lab publications are contrary to the general drift of its content.
>
> These matters, then, stress the need for a radical revision of the ERAB Report. The Great Problem of Fusion – sustainment – has been now achieved under the influence of electric fields in place of high temperatures. It is true that the achievement is halting and as yet unsatisfactory, but the energy density of the best results is that of a nuclear fission reactor – and fusion (i.e., tritium production) has been sustained (in occasional

successful experiments) over many weeks.

I put it to you that the first and necessary step is that the negative ERAB Report, which, in the light of new research findings, no longer corresponds to the new findings, should be formally withdrawn. After four or five years of good research funding may be the time to write a new report.

The remarks of Bockris illustrate the degree of individual subjectivity with which some proponents of cold fusion view the various claims of excess heat and fusion products. In his first paragraph, Bockris takes the liberty to make sweeping generalizations about complex multidimensional institutions as though each had a single homogeneous position on cold fusion. National laboratories, like universities, have hundreds of scientists and only one or two small groups at each institution were claiming confirmatory evidence. It may sound impressive to use institutional names like Los Alamos and Stanford to support a particular view or hypothesis. However, in reality no group at Los Alamos had reported excess heat, two groups had reported a small amount of tritium and one group had reported low levels of neutrons. At Stanford, only a single group had reported small amounts of excess heat from calorimetric measurements. The single group at the University of Minnesota that earlier reported excess heat now obtains an energy balance in its cells. The Bockris statement in the first paragraph focuses on the few scientists who claim some fragmentary evidence for cold fusion, however, he distorts the total picture by ignoring the many scientists at National Laboratories getting negative results.

In the second paragraph, Bockris failed to mention that scientists at the Bhabbha Atomic Research Center (BARC) reported that they produced tritium in a number of their cells during the *first* day of electrolysis. This claim directly contradicts Bockris' induction period of "many weeks of electrolysis." Contrary to Bockris' claim in the fourth paragraph of "direct confirmation of these phenomena from so many sources", the truth of the matter is that recent months have not produced any new confirmatory evidence that meets the standards of peer-reviewed scientific journals. Bockris in his letter alters the meaning of words like confirmation and publication to strengthen his point of view. His statement that the great problem of fusion – sustainment – had been achieved in electrolytic cells with electric fields is a prime example of how far some fusion proponents have deluded themselves. In summary, Bockris' remarks contain over generalizations and misleading statements that serve to suppress the complete truth.

At one time or another there were other instances where strong advocates of cold fusion criticized panel members for a variety of imagined or fabricated shortcomings. A physical electrochemist suggested that "the panel should have principally people whose knowledge and expertise fit the subject area of physical electrochemistry." This was an attempt to question the qualifications of the distinguished electrochemists on our panel by implying they lacked the expertise to judge the electrochemical component of cold fusion. All these panel members had served leading roles in organizations of physical electrochemists as well as having published numerous papers in the field. They were not, however, closely associated with this proponent's special brand of physical electrochemistry, which is related to electrochemical technology. Another proponent referring to two particularly well-known panel members wrote that "neither of those men has expertise in the quantum mechanics of nuclear reaction cross sections at low temperatures while closely confined in metallic lattices." This was an especially ironical charge since this proponent was advocating the appointment to our panel of some well-known theorists who had written papers 'explaining' how the energy from deuterium fusion would couple directly into the palladium lattice without producing high energy gamma rays. The defect in these coupling mechanisms is that they are hypothetical and are not known to occur in nature! Others charged incorrectly that our panel members had an extremely strong vested interest in denying the existence of the Fleischmann-Pons phenomena because of involvement in inertial or magnetic confinement fusion research. This accusation that we were a loaded panel was a scandalous falsehood. And some went so far as to suggest our panel be reconstituted with their favorite choices, whose names they proceeded to give us. These were usually individuals who had gone on record as strong supporters of cold fusion.

Although criticism came from some segments of the sharply divided scientific community, credit has to be given to our panel members, who each worked hard to accumulate information from every possible source and considered this information carefully. Each panel member accumulated a full shelf of material, containing published papers, preprints, formal and informal reports, letters and other forms of communication, all bearing on the cold fusion phenomena. All of our panel's meetings were open to the press and the public. Hence, anyone could have attended the meetings and observed our panel's work first-hand. In my close association with each of the panel members, I saw no evidence of bias on the part of any panel member. It was my experience that panel members bent over backwards to be fair in their attempt to evaluate oftentimes incomplete and contradictory evidence and to come to conclusions based on the information that was available to us.

There is some evidence also that our panel acted fairly and responsibly in fulfilling its charge. On December 13, 1989, S.E. Jones of BYU circulated a letter to the Utah news media, which in large part had been severely critical of our panel and its reports. The letter had particular significance as it came from Jones, the leader of one of the original groups of researchers claiming cold fusion based on their reported trace amounts of neutrons in electrochemical cells. I quote from this letter:

There have been numerous harsh comments lately about the Department of Energy, its cold fusion panel and the co-chairman Professor John Huizenga, quoted in the *Deseret News*, the *Salt Lake Tribune* and the *Provo Daily Herald* and on local television. I wish to present some facts to help round out the picture. People in Utah deserve to know the rest of the story about cold fusion.

University of Rochester Professor John Huizenga was co-chairman of the DOE panel, and has been criticized by Professor Pons by name lately (for example, see *Provo Daily Herald*, Dec. 4, 1989.) I remember how much I appreciated kind words by Professor Haven Bergesen of the University of Utah last spring at a time of much controversy and confusion (for example, see *Deseret News*, April 2, 1989, front page) and now I would like to say a few things in defense of Professor Huizenga. When the DOE panel's draft report came out a few weeks ago, I found some apparent mistakes. Professor Kent Harrison (BYU) and I then sent a letter to Professor Huizenga pointing out the problems. Professor Huizenga called me and we had a very useful conversation. He graciously agreed to correct some of the disputed statements. We disagreed – agreeably – on a few points, as there are points on which reasonable people can disagree.

The Department of Energy set up a panel of scientists to study whether special funding should be provided for cold fusion, such as a national center as proposed by the University of Utah. They concluded that such a center of large funding was unjustified, because experiments "do not present convincing evidence that useful sources of energy will result from the phenomena attributed to cold fusion. In addition, the panel concludes that experiments reported to date do not present convincing evidence to associate the reported anomalous heat with a nuclear process."

Compare that statement to what the previous director (Hugo Rossi) of the University of Utah National Cold Fusion Institute wrote in a report released November 3: "There is as yet no published data from an experiment demonstrating convincingly that the energy balance is positive over its entire course and which ties this excess energy undeniably to a nuclear process." (Published in *Salt Lake Tribune*, Nov. 4, 1989, p. B–2.) The statements by the Department of Energy panel and by Professor Rossi of the University of Utah are in remarkable harmony.

The panel recommended that DOE funding be provided through the normal scientific screening process known as peer review. This is the way cold fusion research has been funded by the Department of Energy at BYU and is the way nearly all scientific research is funded at universities.

Jones performed a service to the press and the people of Utah by comparing an important quote from our report with a statement of Hugo Rossi, the former interim director of the National Cold Fusion Institute. As can be seen in Jones' letter, the two statements are remarkably similar. Only time will tell whether any of our conclusions will need to be modified. As of this time, I am neither aware of any new hard data nor any other information that support the phenomenon known as cold fusion, where large amounts of energy are produced in an electrolytical cell by a nuclear fusion process.

All the evidence available indicated that Fleischmann and Pons initially expected to produce D+D fusion, as was known at that time, where energy and fusion products are generated in well-defined amounts. On the other hand, there is no reported evidence that their original laboratory was equipped with the necessary shielding to protect them from large intensities of fusion products. In their first report [*J. Electroanal. Chem.* **261** 301 (1989)] describing their results on electrochemically driven fusion, Fleischmann and Pons faced a most serious dilemma. Their results were in sharp disagreement with all previous D+D fusion results because their reported excess heat exceeded their reported fusion product yields by over 8 orders of magnitude (10^8). Even so, they proceeded to interpret their data as proof of the occurrence of D+D fusion at room temperature. This was based on Fleischmann and Pons' claim that the excess heat was so large that it ruled out chemical sources. But where were the fusion products? Fleischmann and Pons solved this predicament very easily in their paper by hypothesizing that nearly all of the excess heat they reported was due to an unknown nuclear process. Solving the dilemma of the missing fusion products by the above radical hypothesis opened the way for the irrational conjecture that inside the palladium lattice the well-known properties of the D+D reaction would be severely modified. This in turn led the proponents of cold nuclear fusion to propose and attempt to legitimatize a variety of non-conventional D+D nuclear fusion signals; in particular, uncorrelated excess heat, neutrons, tritium and helium. Once the proponents granted themselves the license to decouple all the conventional correlated signals of D+D fusion, the door was wide open for a myriad of possibilities to confuse backgrounds and mistakes with real signals. This bizarre situation fostered the claims of irreproducibility and sporadicity, well-publicized characteristics of cold fusion.

On examination of all available data at the time, our panel concluded that there is no convincing evidence to associate the reported anomalous heat with a nuclear process and, furthermore, that the present evidence for the discovery of a new nuclear process termed cold fusion is not persuasive. In addition, no known evidence exists to support the extreme conjecture that D+D fusion in a metallic (e.g. palladium or titanium) lattice will have either the required enhanced probability or altered branching ratio to explain cold fusion claims. In all of its deliberations, our panel took the point of view that the proponents of cold fusion have the responsibility to demonstrate convincingly that such a phenomenon exists. Lacking this evidence, our panel concluded that cold fusion had not been established. Nothing has occurred in the intervening months to alter this conclusion. The ardent believers have their backs to the wall and to have any credibility must formulate a well-documented procedure that expert scientists can follow to reproduce their cold fusion claims. Their continuing failure to provide convincing proof of cold fusion will ultimately relegate the phenomenon to the back pages of pathological science.

VIII

Where are the Fusion Products?

Nuclear fusion of hydrogen isotopes has been studied for decades and the resulting fusion products are well known. The fusion of deuterium (D+D), for example, produces neutrons, protons, tritium, the helium isotopes of mass numbers three and four and gamma rays (see the reactions on p. 6). The reported claims of having observed one or more of those fusion products from cold fusion at room temperature are examined in this chapter. At the outset, however, it is worth noting that most scientists do not believe that the large excess heat claimed by Fleischmann and Pons is due to nuclear reactions. First, it is extremely unlikely because of the very high nuclear fusion rate required to account for the claimed large amount of excess heat. The second and more persuasive argument against Fleischmann and Pons' claim is that the commensurate intensities of fusion by-products, the crucial part of the signature of nuclear fusion, have not been found.

Insofar as the proof of whether or not any nuclear fusion was occurring in the palladium lattice during electrolysis depended on the detection of fusion products, searches for these products were given a high priority in laboratories around the world immediately after the University of Utah press conference.

We know a great deal about the nuclear fusion of deuterium (D). It has been studied for over a half century, some of the earliest work dating from Lord Rutherford at Cambridge in the 1930s. The reaction between two low-energy deuterium nuclei proceeds in the three ways shown by reactions (1a), (1b) and (1c) on p. 6. As stated in Chapter I the principal products from low-energy deuterium fusion are neutrons (2.45 MeV) and ^{3}He (0.82 MeV) from reaction (1a) and protons (3.02 MeV) and tritium (1.01 MeV) from reaction (1b). The near equality of the reaction branches (1a) and (1b) has been verified also for muon-catalyzed fusion. In this case the ratio of the product yields from reaction (1a) to reaction (1b) is 1.4 [D.V. Balin *et al.*, *Phys. Lett.* **141B** 173 (1984)]. This value, which is slightly larger than one, is understood theoretically as due to the p-wave character of muon capture in muon-catalyzed fusion [G.M. Hale, *Muon Catalyzed Fusion* **5/6** 227 (1990/91)]. The low probability of reaction (1c) means that the products ^{4}He (0.08 MeV)[15] and high-energy gamma rays (23.8 MeV) have very small yields compared to the fusion products from reactions (1a) and (1b).

Fusion products are the most sensitive signatures of fusion between deuterium nuclei. For reaction (1a), neutrons are the most easily detected product, by direct counting. In addition, the ^{3}He can be measured by sensitive mass spectrometer techniques. For reaction (1b), protons can be counted directly, or the accumulated radioactive tritium can be measured, albeit with lower sensitivity for short experiments due to its 12.3-year halflife. For reaction (1c), the high-energy gamma ray is readily detectable, by direct counting, while the ^{4}He may be identified by mass spectrometer, although the sensitivity is low and contamination from helium in the atmosphere must be eliminated (one cubic centimeter of air contains 1.4×10^{14} atoms of ^{4}He). Information on the energy release associated with each of the three exit channels resulting from the D+D fusion reaction, as well as that for the other hydrogen isotopes are summarized in Table 1.

TABLE 1. Known fusion reactions of hydrogen isotopes.

The approximate branching ratios for the D+D reaction at low energies are included.

Reaction	Energy Release (MeV)	Reactions sec^{-1} per 1 Watt Output	Branching ratio
(1a) D + D → ^{3}He + n	3.27	1.91×10^{12}	~0.5
(1b) D + D → T + p	4.03	1.55×10^{12}	~0.5
(1c) D + D → ^{4}He + gamma	23.85	2.61×10^{11}	~10^{-7}
(2) p + D → ^{3}He + gamma	5.49	1.14×10^{12}	
(3) p + T → ^{4}He + gamma	19.81	3.15×10^{11}	
(4) D + T → ^{4}He + n	17.59	3.55×10^{11}	

The numbers in the third column are the reactions per second per watt of output power for each of the reactions considered individually. For example, the D+D reaction branch (1c) requires 2.61×10^{11} reaction events of this type per second to generate 1 watt of power. Based on the branching ratios in the last column of Table 1, 1 watt of power from the D+D reaction produces 8.55×10^{11} neutrons (and helium-3 atoms) per second, 8.55×10^{11} protons (and tritium atoms) per second and approximately 1.7×10^5 atoms of helium-4 (and 23.8–MeV gammas) per second. As evident from these numbers, 1 watt of excess heat (or, more precisely, excess power) from D+D fusion produces copious amounts of fusion products.

[15] The helium acquires 0.08 MeV of kinetic energy due to the recoil associated with the gamma emission process.

The announcement on March 23, 1989 by Fleischmann and Pons that they had produced watts of power in a test tube by the nuclear fusion of deuterium inside a palladium lattice, caused nuclear scientists to ask almost immediately the critical question: If the excess heat was due to nuclear fusion of deuterium, where are the commensurate numbers of nuclear particles? Had Fleischmann and Pons been exposed to massive amounts of nuclear radiation? Answers were soon forthcoming from the University of Utah scientists themselves. They announced that the intensities of the associated fusion products (neutrons, tritium, etc.) were some billion of times smaller than required to be commensurate with the reported excess heat. In spite of this enormous discrepancy, Fleischmann and Pons were insistent that deuterium fusion was occurring. They recognized, nonetheless, that their reported fusion products were orders of magnitude less than the claimed heat and thus could not be the explanation for the amount of heat. This unaccounted for discrepancy between the reported amounts of excess heat and fusion products, which are by far the most sensitive signatures of fusion, led many early on, to be very skeptical of the University of Utah claims. The claim of a nuclear fusion process suffered another devastating blow soon after the Fleischmann-Pons paper appeared in April, 1989. Scientists from MIT showed conclusively that the Fleischmann-Pons neutron data were incorrect and due to an instrumental artifact, leaving no convincing evidence that any nuclear fusion had occurred (their reported tritium counting rate was within background levels). Hence, at this point in time one might have expected the whole cold fusion episode to blow over.

The fact that the cold fusion episode lingered on beyond the initial two or three months, in spite of the fact that many well-equipped and highly-respected research teams were unsuccessful in their attempts to duplicate the results of Fleischmann and Pons, is a story for sociologists and historians of science. I would surmise, however, that several unusual and sometimes bizarre reports of nuclear phenomena associated with cold fusion experiments fueled scientists' interest in the early months. Among these tantalizing reports were: The early claims made by the Jones group of the detection of very low levels of neutrons during electrolysis experiments, the claims of Scaramuzzi et al. of neutron bursts from non-equilibrium D_2 gas experiments and the more controversial claims of Bockris et al. of tritium production during electrolysis experiments.

The Jones experiments were treated from the start by the scientific community in a very different manner from the University of Utah claims. This was because Jones' reported fusion rate was some thirteen orders of magnitude less than the claimed fusion rate of Fleischmann and Pons based on their reported excess heat, and because, in addition, Jones had stated emphatically that his results had no apparent application to the production of useful energy. In obtaining their reported data, Jones et al. had pushed their experimental techniques to their limit. The resulting neutron signal they obtained was very difficult to isolate from the cosmic ray and other backgrounds. If confirmed, however, these results would be of considerable scientific interest because they implied an enormous enhancement in the fusion rate over theoretical predictions. Hence, it was very important early on to have other experimental groups repeat and attempt to verify the experimental results of Jones et al, which were very unusual in their own right.

In situations where the reported signal is extremely difficult to separate from the associated background, one must worry about unknown sources of error that may be present. Invalid conclusions sometimes result when such experiments are not repeated a sufficient number of times under a tightly controlled set of experimental conditions. Another essential aspect of such experiments is the performance of the proper number of background or blank experiments to give statistical meaning to the final results. Several claims have been made for the production and detection of neutrons at very low levels near background from both D_2O electrolysis and D_2 gas experiments. Most of these claims have been withdrawn when problems with the neutron detectors, especially BF_3 counters, were discovered. Some of the many neutron experiments performed to date, mostly giving only upper limits to the fusion rates, are summarized in Table 2 (p. 141).

Surprises do occur in science. In scientific research it is always important to be on the lookout for an unexpected or surprising result. On the other hand, when one's observations come into direct conflict with long established experimental and theoretical results, it is equally important to look for mistakes and imperative to do a variety of checks, sometimes with alternate procedures. This careful and cautious approach to science is not always practiced. Several cold fusion enthusiasts, for example, have at one time or another when faced with an unexpected result, invoked one or more marvels or miracles that either disregard or directly violate well-founded results and principles of nuclear physics. Since a marvel is something that causes wonder or astonishment and comes as an intense surprise, I'll follow the scheme already introduced in Chapter III and refer to these reported claims as miracles. In order to better understand some of the claims of cold fusion proponents, it is pedagogically useful to amplify here on the description of the different classes of miracles that have been invoked.

1. Fusion-rate miracle

The first miracle invoked by some cold fusion proponents was required to obtain the Fleischmann-Pons reported rate of power production. Positively charged deuterons are held apart by their Coulomb repulsion, and at low energies, fusion occurs by the particles tunneling through a Coulomb barrier. This tunneling probability varies extremely rapidly with the energy of the particles, or with the distance between the deuterons. In D_2 gas molecules, the bond distance is 0.074 nanometers and the theoretical cold fusion rate is 3×10^{-64} sec^{-1} [Koonin and Nauenberg, *Nature* **339** 690 (1989)]. As stated previously this corresponds to one fusion per year for a solar mass of deuterium! If the energy is increased or alternately the distance between the deuterium nuclei is decreased the fusion rate increases rapidly. The separation distance of two deuterium nuclei in a muon-bound molecule, for example, is approximately 207 times smaller than that of ordinary D_2 gas. This short bond distance is the reason for the relatively large fusion rate for muon-catalyzed fusion. In the palladium metallic lattice, however, the mean spacing of the deuterium nuclei is even larger than that in D_2 molecules, ranging in distance from 0.28 to 0.17 nanometers depending on whether the nearest-neighbor deuterium atoms are in octahedral, tetrahedral or mixed sites. The conjecture that the host metallic lattice during electrolysis of D_2O will serve to enhance the fusion rate by more than 50 orders of magnitude (10^{50}) (necessary to sustain Fleischmann and Pons' claim of power production) requires a high-grade miracle [see, for example, calculations of the upper limits of D+D fusion rates in palladium by Leggett and Baym, *Phys. Lett.* **63** 191 (1989) and Wilets *et al.*, *Phys. Rev.* **C41** 2544 (1990)]. Even an enhancement of 40 orders of magnitude (10^{40}), necessary for the Jones *et al.* fusion rate, has still to be included in the miracle class. It so happens that this latter enhancement factor is near the limit of particle detectability with modern counting instruments. Hence, all recorded signals must be critically analyzed to determine whether they are true or background signals. As a result of this ambiguity, a few groups are still reporting positive results whereas a majority of groups have reported their results in terms of an upper limit, the value of which is sometimes one to two orders of magnitude below the value of Jones *et al.*

2. Branching-ratio miracle

Some cold fusion proponents invoked a second miracle in order to alter the well-known branching ratios of the three reaction channels (reactions 1a, 1b and 1c, Table 1, p. 109) associated with D+D fusion at keV energies. These branching ratios are well understood in terms of basic nuclear theory. The nearly equal probability of the reactions $D+D \rightarrow {}^3He+n$ and $D+D \rightarrow T+p$ is the result of the charge independence of nuclear forces. The small probability of the reaction $D+D \rightarrow {}^4He+gamma$ results from the small ratio of the electromagnetic to the nuclear force. One gains considerable confidence that these branching ratios are independent of initial conditions from the observation that they are nearly constant over a range of deuteron energies in the keV region, as well as for muon-catalyzed fusion. All previous experimental evidence supports the position that the effect of the chemical environment on nuclear reactions is essentially negligible except for very small effects associated with reactions involving the atomic electrons. Pressure and chemical effects of very small magnitude have been observed for electron capture and isomeric-transition processes. For example, the halflife of the electron capturing isotope ^{7}Be decreases by about 0.6 percent under a pressure of 270,000 atmospheres [Hensley et al., Science **181** 1164 (1973)]. These effects, however, are understood quantitatively. Extremely minor changes in the D+D reaction branching ratio of the reactions D+D $\rightarrow$ ^{3}He+n and D+D $\rightarrow$ T+p may result from the Oppenheimer-Phillips process described later. Any major modification, however, of the known branching ratios for the D+D reaction in a metallic lattice environment represents a second miracle. In the model of Walling and Simons (see Chapter III), they invoked a radical change in the D+D reaction branching ratios such that only the reaction branch (1c) D+D $\rightarrow$ ^{4}He+gamma ray, occurred in solids. This represents a highest-grade miracle of the second type. Others postulated a branching ratio for the D+D $\rightarrow$ T+p reaction to the D+D $\rightarrow$ ^{3}He+n reaction of 10^6 to 10^9. Such major changes in the latter branching ratio from the well-established value of approximately one cannot be explained in any known way and represents also a miracle of the second type.

3. Concealed-Nuclear-Products Miracle

This third miracle has been invoked by several adherents of cold fusion. Walling and Simons (see Chapter III) assumed the 23.85 MeV of energy associated with the reaction branch D+D $\rightarrow$ ^{4}He+gamma ray was, somehow, miraculously taken up by the lattice, without producing detectable amounts of the high-energy gamma ray. Schwinger appealed to this same miracle in postulating the reaction p+D $\rightarrow$ ^{3}He+energy without detectable amounts of the 5.5–MeV gamma rays. Fleischmann and Pons exercised the ultimate use of this miracle by postulating that essentially all of the excess heat from cold fusion was due to an unknown nuclear process!

In reading the next sections on the reported evidence and claims for observing fusion products from cold fusion, it will be very helpful to keep in mind when one or more of the above miracles are required to rationalize reported results on cold fusion that violate our present understanding of nuclear physics. This will help one in the evaluation of claims and counter claims. In some cases, supplementary data even contradict a proposed miracle. For example, any claim to have produced sizable amounts of tritium through the reaction D+D $\rightarrow$ T+p must be supported by the production of secondary 14–MeV neutrons from the reaction T+D $\rightarrow$ ^{4}He+n. Hence, the miraculous production of tritium without a commensurate yield of these secondary neutrons renders the miracle null and void. This is actually the case for the Bockris and BARC claims for high-level tritium production, where their T/n ratio was reported to be in the range of 10^6 to 10^9.

A. Tritium and Protons

In this section some of the striking claims made of charged particle production in palladium lattices by D_2O electrolysis and D_2 gas experiments will be examined. First, I will focus on the claim that tritium is produced by deuterium fusion via the $D+D \rightarrow T+p$ reaction in electrolytic cells. The first report of tritium production in cold fusion experiments occurred three weeks after the University of Utah's press conference. Two physics graduate students, Messrs. Eden and Liu of the University of Washington, reported a mass spectrometric determination of a tritium-like mass produced in D_2O electrolysis. Their results were reported in the *Wall Street Journal* by Jerry E. Bishop on April 14, 1989. Here is an example where the initial publication by press release, bypassing the standard publication process, eliminated one of the important self-correcting mechanisms in science. These students had made an elementary mistake in the interpretation of their data. It was pointed out to them later in the peer-review process that their claimed evidence was due to triatomic species composed of combinations of deuterium and hydrogen. This illustrates that publication in a peer-reviewed journal is the preferred way to validate and to communicate scientific results.

In late April 1989, tritium was reported to have been found in cold fusion experiments performed in the laboratory of Dr. John Bockris, a long time friend and acquaintance of Fleischmann. The search for tritium initiated one of the most unusual and mysterious stories in the whole cold fusion episode. Soon after the University of Utah press conference, the Bockris group fabricated twenty-four electrolytic cells and began to attempt confirmation of the Fleischmann-Pons claim of excess heat due to a nuclear process. Their palladium cathodes with diameters of one and three millimeters were supplied by Hoover and Strong and those of six millimeters were obtained from SurePure Chemicals, Inc. [*J. Electroanal. Chem.* 270, 451 (1989)]. The source of the cathodes is of interest because of impurities which will be discussed later. The Bockris group had what appeared to be fantastic success! In their first series of experiments, Bockris' student Nigel Packham found high concentrations of tritium in the electrolyte of many of their simple fusion cells.

Packham and his colleagues were startled because it appeared that they had performed the critical experiment proving that the Fleischmann-Pons phenomenon was associated in some way with the D+D fusion reaction.

How else was it possible to have so much tritium in their cells? It is well known that all heavy water (D_2O) contains some tritium, the amount depending on the supplier, the method of enrichment and its history. Furthermore, the tritium in the electrolyte is enriched during the electrolysis. Most reports of excess tritium in cold fusion electrolytic cells can be accounted for by this electrolytic enrichment process. For example, the levels of tritium reported by Fleischmann and Pons are consistent with this interpretation. The Texas A&M results were particularly noteworthy, however, since their tritium concentrations were many orders of magnitude larger than expected from the electrolytic enrichment process. Even so, there was a great disparity between the reported intensities of excess heat and fusion products.

One of the more unusual features of the Bockris group's experiments was that in a short period, six of their electrolytic cells were producing huge amounts of tritium. Kevin Wolf, a nuclear chemist at the Texas A&M Cyclotron Institute, was openly pessimistic about the significance of the Bockris results, however, because all of these cells had cathodes cut from the same strand of one-millimeter palladium wire. By this time, Wolf had begun checking the Bockris cells for neutron emission with his detectors and, in addition, was assisting the Bockris group in assaying the electrolytes of the various electrolytic cells for tritium. Wolf presented the positive tritium results of Bockris at the Santa Fe meeting on May 23–25, 1989, and while there, was contacted by a member of the Bockris group informing him that a 3-millimeter cathode from Hoover and Strong was also producing tritium. The participants of the Santa Fe meeting had serious concerns about the origin of the tritium in the Bockris cells. The data were simply that remarkable. Wolf was known to be an experienced and careful experimenter, hence, most agreed that the cells probably contained tritium, but where was the tritium coming from? The answer to this question was particularly puzzling because other independent groups doing cold fusion experiments at Texas A&M University were not able to produce tritium. For example, John Appleby, who reported excess heat at Santa Fe but no tritium, asked Bockris the question on the minds of many of the conferees, "Are you sure that somebody hasn't been spiking your cells?"

The extraordinary claim of tritium production was the main topic on the agenda during the visit of the DOE/ERAB panel to Texas A&M University in June, 1989. Nigel Packham, a fifth year graduate student in the Bockris laboratory presented the tritium data. One of the Bockris cells was of particular interest because it was claimed that the buildup of tritium as a function of time had been observed during electrolysis carried out at high current density. The reported tritium activity was 5.2×10^3, 5.0×10^5 and 7.6×10^5 disintegrations per minute per milliliter of electrolyte, two hours, six hours and twelve hours, respectively, after the initiation of the high current density [*J. Electroanal. Chem.* **270** 451 (1989)]. The initial activity was 1.0×10^2 in the above units. Packham presented these data to our panel by fitting a smooth curve through the four points! Jacob Bigeleisen, one of the DOE/ERAB panel members whose experience with tritium dates back to the Manhattan Project, pointed out that the four points did not uniquely define their smooth curve. He pointed out that a simple step function graph rising at six hours and remaining flat could equally well be drawn through the data. To this Kevin Wolf said, "Jake are you implying that someone spiked that sample?". Bigeleisen said, "Kevin, you said that, I would never say such a thing."

The first manuscript by the Bockris group describing the production of tritium from D_2O electrolysis at a palladium electrode was submitted to *Nature*. However, it was rejected as the result of the peer-review process due to inadequate documentation and lack of controls. Different journals responded to the profusion of manuscripts on cold fusion at this time with different degrees of scrutiny. *Nature* subjected all the papers it received on the subject of cold fusion to a peer-review process with standards expected for a leading scientific journal. The same cannot be said for a number of other less prestigious journals. For example, after being rejected by *Nature*, the manuscript of Bockris was then submitted to, and accepted by, the *J. Electroanal. Chem.*, **270**, 451 (1989). This publication described the observation of tritium produced in eleven D_2O electrolysis cells, nine of which had very high levels of tritium. Of these, eight cells had one-millimeter cathodes and one cell had a three-millimeter cathode, all of which were supplied by Hoover and Strong. Bockris in a preprint of this manuscript raised the possibility that the high tritium concentrations were due to "hidden or secret interference." However, he dismissed the possibility of sabotage in the published manuscript with the following argument, "Interference with the experiments is considered improbable because of positive results from the Cyclotron Institute to which entrance is prohibited except by the usual personnel at the Institute." Having visited the Cyclotron Institute many times, I found the Bockris statement unconvincing. Many people have keys to the Institute and spiking, if it had occurred, would have been easy to do and very difficult to detect. Furthermore, only two of the eleven cells in the Bockris manuscript were run at the Cyclotron Institute.

Following the early series of experiments described above, the Bockris group was less successful in producing tritium. At the First Annual Conference on Cold Fusion, Bockris reported that fifteen out of a total of fifty-three cells had produced tritium, corresponding to only four positive results beyond the eleven positive results reported in the early manuscript. Nine of the thirteen cells with one-millimeter cathodes gave positive results. These data disagree slightly with the compilation of Wolf *et al.* (NSF–EPRI Workshop, October 16–18, 1989) where eleven of the one-millimeter cathodes gave positive results. Low level neutrons, at the approximate level of one neutron per second, were observed by Wolf for a single cell giving positive tritium results. Although this value is more than an order of magnitude larger than the neutron level reported by Jones *et al.* it is many orders of magnitude less than the reported production of tritium. The absence of commensurate levels of neutrons, either primary or secondary, suggested to most scientists that the tritium had not been produced by the $D+D \rightarrow T+p$ reaction in the cell but had entered through some type of contamination. Bockris, however, interpreted the Texas A&M tritium results as certain. On July 18, 1989, he wrote to the DOE Panel:

> Now, in the experiments carried out at Texas A&M, it is possible to say that the production of tritium was certain. I base this unusual statement on the fact that the liquids from these results were verified by five independent laboratories . . . Now it is up to the panel, of course, to explain the method of producing tritium from deuterium which does not involve fusion. Personally, I cannot conceive a *chemical* way of converting one isotope to the other. Electrochemical confinement has attained the goal of fusion physics: sustainment.

The fact that five independent laboratories confirmed the presence of tritium in the Bockris' cells in no way proved that the tritium was produced by a nuclear reaction. The same outcome would have resulted if the origin of the tritium were due to contamination. When taking his sharp and sarcastic poke at the panel requesting that they explain the method of producing tritium from deuterium without involving fusion, Bockris once again attempted to divert attention away from the horrible possibility of contamination.

In a later letter (10/26/89), Bockris reemphasized "I want to point out to you that as far as a straightforward matter of 'Do nuclear reactions take part at electrodes?', there is absolutely no doubt whatsoever that this occurs." The panel members at no time doubted the presence of tritium in the Texas A&M cells, largely because of Kevin Wolf's involvement in confirming its presence. The important unanswered question, however was, where is the tritium coming from? The panel was simply unwilling to accept Bockris' claim of tritium as absolute confirmation that nuclear reactions were occurring in electrochemical cells. The likelihood that the tritium was due to contamination was far too great to be lightly dismissed. It was well known, for example, that the Bockris laboratory contained a strong source of tritium, as do many other chemical laboratories.

Several experimental groups had reported small amounts of tritium that could be accounted for by tritium enrichment during the electrolysis process. The most extensive and systematic searches for tritium in the electrolysis of D_2O with palladium cathodes had been carried out by the groups of

Charles Martin at Texas A&M, Edmund Storms and Carol Talcott at the Los Alamos National Laboratory, and, more recently, by Kevin L. Wolf *et al.* of Texas A&M. The Martin group had run over eighty cells, many with advice, materials and protocal from the Bockris group, and none of them produced any tritium above that expected from the isotopic enrichment during electrolysis. On the other hand, Storms and Talcott claimed a small amount of tritium production in some 13 out of a total of 150 cells (*Proc. of the First Annual Conference on Cold Fusion*, page 149 (March 28–31, 1990). However, these measurements were seriously flawed because only some six control cells were run. Following his analysis of the Storms and Talcott data, Bigeleisen wrote to Storms (November 10, 1989) and stated:

> I do not find convincing evidence in the data you submit to substantiate your claim that excess tritium is being produced in your cells. Nor do your data disprove it. What is clear is that your data are irreproducible and show scatter much larger than can be ascribed to counting statistical errors.

One of Storms' and Talcott's cells had 2500 disintegrations of tritium per minute per milliliter in a spill but only about 100 disintegrations per minute per milliliter in the cell. Another cell had 3000 disintegrations of tritium per milliliter of electrolyte, however, this cell was never measured for tritium before this level was achieved. The Storms and Talcott cells giving excess tritium have amounts that are orders of magnitude smaller than those of the Bockris group, and for reasons mentioned above do not give convincing evidence that the tritium is being made by a nuclear reaction.

All of this work on tritium was well summarized by Wolf in a paper presented at the NSF/EPRI Workshop in Washington (October 16–18, 1989). He commented on both his attempts to reproduce the Bockris group's tritium measurements and his attempts to establish whether the source of tritium was from contamination or nuclear reaction. The abstract of Wolf's paper stated:

> The sudden appearance of tritium activity in the cells requires the tritium to be loaded in a component prior to the beginning of cell operation in a contamination model. Release is assumed to be caused by deterioration of one of the materials used in the 0.1 M LiOD solution. In an extensive set of tests, no contamination has been found in the starting materials or in normal water blanks. Results for neutron and gamma-ray correlations have proved to be negative also. The limit set on the absence of 2.5–MeV neutrons for the T/n ratio is 10^7 from that expected from the D+D reaction, and 10^3 for 14–MeV neutrons expected from the T+D secondary reaction. Similarly, Coulomb excitation gamma rays expected from the interaction of 3–MeV protons with palladium are found to be absent, which indicates that the D+D $\rightarrow$ T+p two-body reaction does not occur in the palladium electrode.

Wolf here concurred with the analysis of our DOE Panel that none of the tritium found in any of the Bockris experiments was produced by the known D+D reaction. As of October 1989, however, Wolf *et al.* had not identified the source of the tritium contamination.

The above conclusion that tritium was not produced by D_2O electrolysis is in agreement also with experiments which searched directly for the 3–MeV protons produced in the $D+D \rightarrow T+p$ reaction. A variety of experimental techniques had been used in these searches for protons; all of these studies set very low limits of fusion occurring via the $D+D \rightarrow T+p$ reaction. For example, Price *et al.* [*Phys. Rev. Lett.*, **63** 1926 (1989)] set an upper limit of 8×10^{-26} fusions per DD pair per second for the $D+D \rightarrow T+p$ reaction. These direct searches for charged particles confirms that the tritium activity reported by Bockris, and others was not due to D+D fusion. The Price *et al.* limit on the cold fusion rate was also considerably less than the fusion rate inferred from the neutron measurements of Jones *et al.* via the $D+D \rightarrow {}^3He+n$ reaction.

The only other laboratory in the world that had reported large amounts of tritium comparable to that of the Bockris group was the Bhabha Atomic Research Center (BARC) in Trombay (near Bombay), India. At BARC, a wide variety of experiments were carried out by twelve independent teams of scientists employing both electrolytic and gas phase loading of deuterium in palladium and titanium metals to study cold fusion. These results were published in a long report BARC–1500 (December, 1989) which was reproduced in *Fusion Technology*, **18** 32–94 (1990), and also in abbreviated form, in a paper published in the *Proceedings of the First Annual Conference on Cold Fusion*, page 62 (March 28–31, 1990). Our DOE/ERAB panel had these results in the form of a preprint during the time we were preparing our final report.

The BARC experiments were limited to the detection of fusion products with the aim of verifying the nuclear origin of cold fusion, rather than the measurement of "excess heat" since they stated that the heat measurements required intricate calorimetry. This expressed opinion on calorimetry was a striking admission when compared to many early pronouncements describing the experiment as so simple that it could be done in a high school laboratory. One of the unique features of the BARC experiments was that bursts of neutrons and tritium were reported to have occurred in eight out of eleven cells on the very first day of initiating electrolysis! Compare this to the view of Bockris expressed in a letter to me where he stated:

> As to those expert and respected scientists (e.g., at Yale, MIT, Caltech, etc.) who denied the existence of nuclear electrochemical effects, it is easy to understand the absence of their results: the effects, now seen by so many, do not switch on for many weeks of electrolysis.

Pons and Fleischmann had made similar statements, namely:

> Our calorimetric measurements of the palladium-deuterium system . . .
> showed that it is necessary to make measurements on a large number of
> electrodes for long times (the mean time chosen for a measurement cycle
> has been three months).

Hence, in contrast to other experimenters, the different BARC groups were
unusually successful, reporting significant amounts of both neutrons and
tritium from a large variety of different types of cell configurations in times
as short as a day, compared to months for Bockris, Pons and Fleischmann.
The above statements of these authors are, however, inconsistent with some
of their earlier reports. The Bockris group, for example, was also very
successful in producing tritium in their early experiments as described previously. It was later that their success rate decreased markedly.

A second striking feature of the BARC groups' results was that the
measured neutron to tritium yield ratios in eight out of eleven doubly
successful experiments was in the range of 10^{-6} to 10^{-9}. Even neglecting
reaction (1a) shown on p. 6, which is the primary source of neutrons, such
small neutron to tritium ratios leads to a serious inconsistency. The tritons
(T) produced in the D+D reaction have an energy of 1.01 MeV. These
tritons bombard the deuterium in the cathode, as well as the deuterium in
the surrounding heavy water (D_2O) in the electrolytic cell. The reaction
$T+D \rightarrow {}^4He+n$ is a copious source of 14–MeV secondary neutrons. For
tritons stopped in the deuterium loaded in the palladium cathode, the
secondary neutron yield is approximately 2×10^{-5} neutrons per triton, while
for tritons stopping in heavy water the yield is approximately 9×10^{-5} neutrons per triton (U.S. Department of Energy Report DOE/S–0073, November, 1989). Hence, the above neutron to tritium ratios reported by the
BARC scientists were some 20 to 20,000 times smaller than expected from
the secondary neutrons alone!

The reported tritium/neutron (T/n) ratio in the range of 10^6 to 10^9 (note
that the ratio reported here is inverted from that in the above paragraph)
was in direct conflict with the well-established ratio of approximately one
and led cold fusion proponents to invoke miracle number two, the branching-ratio miracle. In this particular case, however, the proponents appeal to
miracle two was pointless because the experimental absence of secondary
neutrons ruled out the possibility that the reported high yields of tritium
were due to the reaction $D+D \rightarrow T+p$. Most nuclear scientists, therefore,
concluded that the tritium measured by the BARC scientists in the above
eight cells was due to some form of contamination.

A third unusual feature of the BARC experiments was the success rate of
their experiments which employed a large variety of electrolytic cells and
electrolytes. For example, all five experiments using NaOD electrolyte (in

contrast to the usual LiOD) produced neutrons and tritium after a short period of electrical charging. This was in sharp contrast with findings of U.S. groups who reported failures with NaOD. Relative to the many tens of groups that had searched for fusion products the world over, the claimed success rate of the BARC groups for producing both tritium and neutrons was nothing less than sensational.

A fourth feature of the BARC experiments was the magnitude of the tritium and neutron production. While the tritium levels are higher than any group in the world except the Bockris group, with which they are comparable, the BARC neutron production rate exceeded by more than a factor of 100 the rate reported by Jones *et al.*

In summary, when compared to experimental results from around the world, the BARC results were too good to be believable. They were neither consistent with the many previous negative results nor with previous positive results. Their reported very large ratios (10^6 to 10^9) of tritium to neutrons served as evidence that the tritium in their cold fusion experiments was not being produced by deuterium fusion. A final comment on the BARC results may be of interest. Even if one assumed the tritium in the BARC cells was due to deuterium fusion, the resulting associated energy was so many orders of magnitude less than the Fleischmann-Pons reported excess heat, that by no stretch of one's imagination could a causal connection be made between the two phenomena.

Recently, Kevin Wolf has begun to unravel the mystery surrounding the tritium at Texas A&M. In his talk at the NSF/EPRI meeting, he had already concluded that there was no evidence for the $D+D \rightarrow T+p$ two-body reaction occurring in the palladium electrode during electrolysis and hence, the tritium did not arise from cold fusion. This, however, still left open the question about the origin of the tritium. In a partial answer to this question, Wolf found that some of his palladium electrodes, purchased from Hoover and Strong, were contaminated with tritium prior to arrival in his laboratory. He concluded "It's pretty clear that our low-level tritium was due to contamination" (*Science*, **248** 1301 (1990)]. These results throw a shadow of doubt on some of the reported tritium work. Recall that the one- and three-millimeter cathodes giving tritium in Bockris' laboratory were also from Hoover and Strong. The Bockris cells, however, showed tritium levels much higher than that expected on the basis of the contamination found by Wolf. In trying to solve the origin of the high tritium levels in the Bockris cells, Wolf made a very astute observation. On analysis of the electrolyte from a Bockris cell that had previously shown high tritium, he found that the cell contained a large amount of light water (H_2O). This finding, although not proof, was consistent with the cell having been spiked with tritiated water which contains mainly light water. This particular cell had been stored for some months in a sealed container and, unless contaminated, should not have contained the observed amounts of light water. After learning of Wolf's results, the Bockris group examined more of their cells and found again in some other cells large amounts of light water. Wolf responded "It's just incredible, I don't understand it." He found less than 1% light water in the cells in his own laboratory. Again the question, was tritiated water, known to be stored in the laboratory, the source of the light water? Wolf said "The proper conclusion is that things in the Bockris laboratory were so uncontrolled and so sloppy that those studies don't mean anything" [*Science*, **248** 1301 (1990)].

In a recent abstract submitted to the BYU meeting on Anomalous Nuclear Effects in Deuterium/Solid Systems (October 22–24, 1990), Wolf *et al.* were very forthright in their statement that tritium was not being produced by D_2O electrolysis. The authors stated:

> A test of the reproducibility of tritium production in D_2O electrolysis with Pd–Ni–LiOD cells has proved to be negative. The results of Packham *et al.* [*J. Electroanal. Chem* **270** 451 (1989)] are considered to be spurious, and are attributed to tritium contamination. An extensive study with over 100 electrolytic cells has shown that the frequency of occurrence of tritium sightings is explained by the tritium contamination found in the palladium metal stock. A cold fusion mechanism is not supported and implications are discussed for many reported tritium sightings in other studies which were conducted without sufficient blank and control experiments.

Neither the Martin nor the Wolf groups, both at Texas A&M University, observed any electrochemically produced tritium in their respective large series of experiments run under well-controlled conditions. These authors have struck a death blow to the often repeated reports by the Bockris-Packham group of large amounts of tritium in their cold fusion cells.

The evidence supporting the conclusion that the high levels of tritium reported by the BARC and Bockris groups is *not* nuclear in origin (i.e. not coming from the $D+D \rightarrow T+p$ reaction) is extremely persuasive and summarized below.

(1) Direct searches for the three-MeV protons from the $D+D \rightarrow T+p$ reaction were all negative, with upper limits orders of magnitude below the reported tritium intensity. One experiment pushed the upper limit for protons to a value much smaller than the neutron yield reported by Jones *et al.* for the $D+D \rightarrow {}^3He+n$ reaction.

(2) The reported neutron to tritium (n/T) branching ratios of 10^{-6} to 10^{-9} disagree by factors of 10^6 to 10^9 with measurements that showed that the ${}^3He+n$ and $T+p$ branches [see reactions (1a) and (1b) on p. 6] from

D+D fusion are approximately equal at energies of a few keV. Calculations confirmed that the magnitude of the Oppenheimer-Phillips effect is less than a few percent for D targets bombarded with low energy deuterons and that this process is irrelevant for enhancing the T+p branch in cold fusion.

(3) The reported neutron to tritium (n/T) branching ratios of 10^{-6} to 10^{-9} are incompatible with the D+D $\rightarrow$ T+p reaction occurring within a palladium cathode surrounded by D_2O because of the secondary neutrons (14–MeV) that are generated by the 1.0l MeV tritons (D+T $\rightarrow$ ^{4}He+n).

(4) Direct searches for the Coulomb-excitation gamma rays expected from the interaction of the three-MeV protons (see reaction 1b on p. 6) with palladium have all been negative.

(5) Tritium impurity has been found by Kevin Wolf in palladium purchased from Hoover and Strong.

(6) None of the approximately 200 cells which were run under well-controlled conditions by the Martin and Wolf groups, colleagues of Bockris at Texas A&M University, showed any nuclear-produced tritium.

In spite of the negative evidence, Bockris *et al.* in a recent paper [*Fusion Technology* **18** 11 (1990)], set forth what they considered to be the two main characteristics of cold fusion. They stated "One is the large tritium to neutron ratio, on the order of 10^8 and the other is the sporadicity and irreproducibility of the phenomena." Such a characterization of cold fusion will, I am sure, surprise almost everyone, friend and foe of the phenomenon. Bockris *et al.* went on to say "A suitable cold fusion theory or model *must* explain both of these two features." Bockris' dislike for conventional nuclear physics and nuclear theory have been stated previously. He wrote to the DOE/ERAB Panel on June 12, 1989 and stated that:

> We are particularly unenthusiastic in the discussion of the application of present theories of fusion in plasmas to the idea of fusion in electrochemical confinement because we think that the difference of conditions, particularly in respect to screening by electrons of the deuterium deuterium interactions is an extreme one . . . Historically, when new science is emerging, it is often reviled and denigrated until the new paradigm is accepted . . . We think that, in attempts to verify a newly claimed phenomena, negative results have much less value than positive ones. Negative results can be obtained without skill and experience.

I find Bockris' statement about negative results, which were obtained by a large number of outstanding research groups working on cold fusion, one of the most eccentric evaluations of experimental results that I've ever heard of. There is ample evidence, however, that Bockris' strange attitude on negative results is held by other proponents of cold fusion also. During the spring meeting of the Electrochemical Society (5/8/89) in Los Angeles, a call for papers issued for a session on cold fusion, specified that only papers confirming cold fusion would be accepted. The organizers justified this decision on the grounds that "since the session is on cold fusion, research that doesn't find fusion would not be relevant." The same restricted selection of data was attempted at the Santa Fe Workshop when an evening session on May 23, 1989 was announced requesting only those papers with positive results. A rebellion by a small group of people resulted in the acceptance of a negative paper. Bockris in the above paragraph also displays his strong belief that the chemical environment will alter the well-known nuclear properties of the D+D fusion reaction.

As one might have expected, out of the smorgasbord of models listed by Bockris *et al.*, only their model "can effectively explain both the sporadicity and irreproducibility of the cold fusion experiments . . . and also the large tritium to neutron ratio." How is it possible for any model to explain or account for these assigned most unusual characteristics of cold fusion? First, consider why the proponents of cold fusion have placed so much importance on tritium. David Worledge of the Electric Power Research Institute (EPRI) stated the case for tritium clearly. He wrote "We shall see that the evidence on tritium generation is the strongest of the three types of evidence for cold nuclear reactions, i.e., tritium, neutrons and heat. It seems no longer reasonable to assume these results are necessarily wrong solely because of theoretical improbability based on current understanding" (*Proc. of the First Annual Conference on Cold Fusion*, page 252, 1989). This is a striking case where the observation of the BARC and Bockris groups came into direct conflict with the long-established branching ratio of approximately unity for the yield ratio of the T+p and ^{3}He+n branches from the D+D reaction, as measured down to deuteron energies of a few kilo-electron volts (keV). Established nuclear theory predicted that this branching ratio would change by only a small amount when the deuteron energy is reduced to thermal energies. Even though no other research groups in the world had been able to produce these high levels of tritium in their experiments and the BARC and Bockris tritium to neutron branching ratios were more than a million (Bockris stated on the order of 10^8) times larger than predicted on the basis of established experiments and nuclear theory, the proponents of cold fusion chose to accept miracle two, and to make this large ratio one of the characteristics of cold fusion. In addition, this large ratio was internally inconsistent due to the absence of secondary neutrons that would result from the tritium interacting with the deuterium in and around the palladium cathode. Therefore, it was ironic that the large tritium to neutron ratio claimed by the proponents to be an important characteristic of cold fusion, actually served to prove that the tritium was not coming from the D+D fusion reaction! Still Worledge of EPRI (an organization that funded cold fusion research) thought that tritium was the strongest evidence for cold fusion.

On what basis did the proponents of cold deuterium fusion justify the large branching ratio in favor of the T+p branch over the ^{3}He +n branch (see reactions (1b) and (1a) on p. 6)? Here the proponents relied on *conventional* nuclear physics theory, in particular, the well-known Oppenheimer-Phillips process. Oppenheimer and Phillips [*Phys. Rev.* **48** 500 (1938)] noted over fifty years ago that the Coulomb field of the target nucleus acts only on the proton in the deuteron and not on the deuteron's center of mass. This leads to an effective polarization of the deuteron, and a mechanism for enhancing the D+D $\rightarrow$ T+p reaction relative to the D+D $\rightarrow$ ^{3}He+n reaction, in the direction required by the BARC and Bockris reported data. Crucial, however, is the magnitude of the Oppenheimer-Phillips effect and its variation with energy, particularly at the very small energies of interest here. Koonin and Mukerjee [*Phys. Rev.* C42 1639 (1990)] found that the effect is less than a few percent and irrelevant to low-energy D+D reactions. Recall that the claims of cold fusion proponents at BARC and Texas A&M require an enhancement in the T/n branching ratio of one hundred million. For such a result, miracle two is indispensable.

And what about the other main characteristic of cold fusion, namely that it is sporadic and irreproducible? The Bockris group had suggested that fusion occurred on the cathode surface where dendrites (whisker-like projections) grow during prolonged electrolysis on the electrode surface. As evidence of a surface model, they appealed to palladium isotope ratio changes during electrolysis, data that was presented at the NSF/EPRI Workshop, but now known to be incorrect (*Proc. of the First Annual Conference on Cold Fusion*, p. 272, 1990). Reproducibility is an important factor in science, especially so when insufficient blanks and controls have been run and all the associated questions have not been completely resolved. Hence, to have as one of two main characteristics of cold fusion its irreproducibility is to put the phenomenon on a subjective, non-scientific basis. To be accepted as an established phenomenon, an experiment in physical science must be reproducible. A set of instructions must be available to allow a competent and suitably-equipped scientist to perform the experiments and obtain essentially the same result.

It is not at all unusual for pioneering experiments to be affected by poorly understood factors. However, when there is a real, new phenomenon involved, this variability can be shown to others. For example, in the early days of semiconductor science, it was easy to demonstrate that some germanium crystals were better electrical conductors than others. The fact was indisputable, since the pieces were in hand and their resistance could be measured again and again. For real, new phenomena, the sources of variability in an experiment are systematically discovered and brought under control; for example, by the control of trace impurities in semiconductors. There has been no sign of this growth of understanding of cold fusion either in the production of fusion products or excess heat.

Dramatic new effects in laboratory science often meet a skeptical scientific community. This is as it should be since systematic skepticism is very critical in science and serves to minimize individual subjectivity. Whether new findings are ultimately accepted depends not on how closely they conform to the prevailing scientific dogma but on whether they can be reproduced at will anywhere in the world. Pasteur's discovery of right and left-handed molecules and Rayleigh and Ramsay's discovery of the rare gases are well-known discoveries which, by the overwhelming force of reproducible experiments, overcame bitter resistance in the scientific community. The few research groups that have claimed to produce large amounts of tritium have not demonstrated it to scientifically competent outside observers. In fact all of the available evidence indicated that the high levels of tritium claimed by the BARC and Bockris groups were not due to deuterium fusion via the reaction $D+D \rightarrow T+p$.

As Wolf *et al.* have stated, many of the cold fusion experiments reporting tritium have been carried out without sufficient blank and control experiments. In an effort to establish credibility for cold fusion, most proponents are much more interested in counting the numbers of groups claiming positive results than examining whether their claims have any validity. For example, in an effort to nullify an outstanding paper reporting negative results, the editor of *Fusion Facts* stated ". . . that over ninety scientists have replicated cold fusion . . ." (*Fusion Facts* 2, no. 4, October (1990), p. 30). In a September (1990) preprint (lecture at the World Hydrogen Energy Conference in Honolulu, July 24, 1990) entitled, "Is there evidence for fusion under solid state confinement?", Bockris claimed that seventy-nine groups in sixty laboratories in twelve countries have been successful in reproducing cold fusion. Anyone familiar with the field can strike case after case from these lists, because either the authors have retracted their original claim or the claim is unsubstantiated and not reproducible. A single well-researched and careful experiment with sufficient blank and control experiments giving a consistent set of positive results would be much more impressive. One such result, however, with excess heat and a commensurate amount of fusion products still does not exist!

In view of the above conclusive evidence that the high amounts of tritium were not coming from the $D+D \rightarrow T+p$ reaction, the most likely source is contamination. However, the Bockris' levels of tritium greatly exceed the level of contamination in palladium found by Wolf. This suggests that the Bockris cells were either run under slipshod conditions facilitating high-level tritium contamination or the tritium was knowingly introduced. Bockris raised the latter possibility himself in his first paper on tritium [J. of Electroanal. Chem **270** 451 (1989)], but went on to argue that it wasn't likely (see previous discussion). The recent discovery by Wolf that some Bockris cells rich in tritium contain large amounts of light water has raised the issue of possible fraud and brought the discussion out in the open (Science **248** 1299, 1990). The presence of light water is consistent with the hypothesis that the Bockris cells were spiked with tritiated water, however, does not prove it. Fraud has often been defined such as to encompass a wide spectrum of behaviors. It can range from selecting only those data that support a hypothesis while concealing other data to outright fabrication of results. The incidence of fraud in physical science is usually assumed to be low, because these experiments are reproducible and can usually be verified in a short time, in contrast to some areas of the life sciences. The characterization of cold fusion as irreproducible leaves it vulnerable to the possible charge of fraud. The effects of fraud on other scientists and the public can be devastating. However, as other investigators skeptically review and attempt to verify previously reported data, fraudulent data and hypothesis are eventually uncovered. Valuable time and research funds are wasted in setting the scientific record straight.

Serious questions had been raised about the high levels of tritium in the Bockris cells from the beginning. Suspicions are bound to be raised when the reported data violate established nuclear physics and large amounts of tritium cannot be reproduced by any experimental group in the world except experimenters at BARC (and here under very different conditions). How was it possible that these two groups could produce large amounts of tritium, within short times after initiating cold fusion experiments, when no other groups in the world have been able, up to this time, to reproduce their miraculous results? In view of the cloud of suspicion that hung over the Bockris experiments, he did little to insure that his cells were adequately protected from possible interference. At one point Bockris removed a fifth-year graduate student from the tritium work, only to have him reappear some three months later with yet two more cells containing tritium! Others in the Bockris laboratory relayed their suspicions about the source of tritium to Bockris only to be rebuffed by Bockris with the declaration that a number of groups had verified his work. The only recourse for unhappy researchers was to leave the Bockris laboratory [Science **248** 1299 (1990)].

The Texas A&M administrators at all levels were reluctant to investigate what was occurring in the Bockris laboratory. Professor Charles Martin, a colleague of Bockris in the Texas A&M Chemistry Department went to Dr. Michael B. Hall, Head of the Chemistry Department and voiced his suspicions. "I warned Hall that I thought there was a very good chance the experimental results were the result of fraud" (Science **248** 1299, 1990).

Martin was one of the principals in cold fusion research from the beginning and had examined many cells in his own laboratory. At one point he had even offered to run Bockris' cells in his own laboratory. Martin wrote in January, 1990 to Dr. John P. Fackler, Jr., Dean of the College of Science, about his own results, "*none* of the eighty-three cells which were run by my students in my laboratory have produced tritium levels above those predicted by the (known) separation factor" (*Science, ibid.*). On any inquiry by the Texas A&M administration, Bockris would retort that a new field of nuclear electrochemistry was in the making with the characteristics of sporadicity and irreproducibility. Following his early startling successes, even he himself would have to wait some ten weeks after starting the electrolysis before fusion commenced (recall the BARC experiments were successful in producing large amounts of tritium in eight out of eleven cells on the first day of electrolysis!). Under the aegis of irreproducibility, cold fusion claims were given some degree of respectability and the administration chose not to initiate an inquiry. Dr. Robert L. Park, Executive Director of the Office of Public Affairs for the American Physical Society, in a July 12, 1990 speech to the Committee on Science, Engineering and Public Policy (COSEPUP) Panel said:

> The Texas A&M administration, which for months had steadfastly refused to investigate allegations of tampering in the cold fusion laboratories, responded by criticizing *Science* for airing the controversy. The reaction of A&M officials was all too predictable to those who have followed the reluctant responses of universities to charges of conflict of interest and fraud . . . But the response of university officials is understandable. Instances of deliberate fraud, thankfully are rare. Far more frequent are the occasions on which a university must defend the rights of faculty members to hold unpopular or nonconformist views.

The Bockris claims of producing high levels of tritium illustrates the problems associated in dealing with possible fraud. Here is a case where all the evidence argued against the production of large amounts of tritium in electrolytic cells, yet no action was taken for many months by the Texas A&M University administrators. Universities must have a policy in place that protects academic freedom, on the one hand, but allows low-level investigation in suspicious cases without requiring a full-blown trial with a formal accuser. The threshold for the latter process is usually too great for it to be timely and effective.

The Texas A&M administration did eventually compose a local panel of three, Professors Fry, Natowitz and Poston, to investigate the Bockris reports of production of high levels of tritium in his group's cold fusion cells. Although the panel commented on most of the cold fusion research at Texas A&M, Bockris' findings were emphasized. The Texas A&M panel's report was basically a review of well-known claims and reports from Texas A&M and contained no surprises. It stated that the Bockris group had obtained very high levels of tritium early on in April and May of 1989, and then again in November. Since then no additional cells were reported to have produced unusual amounts of tritium. The Texas A&M panel's statement that tritium measurements are not now a priority of the Bockris group is a revealing admission. How is it possible not to give tritium the highest priority when Bockris and other proponents of cold fusion have advertised repeatedly that tritium production is the strongest evidence for cold fusion? At the First Annual Conference on Cold Fusion the believers had suggested that the production of tritium was the "unassailable signature of a nuclear reaction."

The Texas A&M panel concluded that the very high levels of tritium reported by the Bockris group at particular times, the surprising amounts of light water present in these Bockris cells, the lack of security in Bockris' laboratory, etc. did not provide convincing evidence for fraud. This is not a surprising conclusion for a local panel. Proof of fraud is very difficult to establish and requires unconditional evidence. It is much easier, in the words of Wolf quoted earlier, to assign the whole tritium episode at Texas A&M to sloppy and uncontrolled conditions in the Bockris laboratory.

Since the Bockris group has not produced unusual tritium levels in their cold fusion cells for over a year, and their new admission that tritium is no longer a priority, one expects that the whole tritium episode, once claimed to be the best evidence for cold fusion, will gradually fade into obscurity.

B. Helium

The Utah helium saga took many odd twists and turns during the weeks and months following the March 23, 1989 press conference. From the beginning some proponents held up the searches for helium as the potential proof of cold fusion. Fleischmann and Pons, for example, acknowledged openly that helium was the critical test. This charged atmosphere led to early claims, based on very shoddy experiments, of the production of the helium isotope of mass number four (^4He) by electrolysis of D_2O with a palladium cathode. These claims, which turned out to be false, in turn triggered some of the most bizarre 'theories' that one could imagine for the purpose of explaining cold fusion. These 'theories' were especially attractive from the point of view of heat and energy production, since each D+D fusion event produced 23.8 MeV of heat (see reaction (1c) on p. 6]) by a radiationless relaxation process where the entire energy was transferred to the palladium lattice. One couldn't wish for a more attractive source of fusion energy, free from the normal much more copious sources of neutrons (reaction (1a) on p. 6) and tritium (reaction (1b) on p. 6) from the D+D reaction.

The helium story had its beginning in the original Fleischmann-Pons press conference. These authors reported an amount of heat over a hundred million times greater than that of the reported fusion products. This enormous discrepancy led Fleischmann and Pons to state in their initial publication that the heat was due to an hitherto unknown nuclear process or processes. How Fleischmann and Pons decided to go public with a declaration of a cold nuclear fusion process with such a gigantic discrepancy between the claimed heat and fusion products is a question that historians of science will ask and debate for years. Somehow they were completely out of touch with the whole field of nuclear physics and gave little attention to the body of scientific knowledge gathered in this field over the last fifty years. Within three weeks of the historic press conference Pons and Hawkins informed Walling and Simons [J. Phys. Chem. 93 4693 (1989)] that they were producing large amounts of ^{4}He in their cells. The claim was that mass spectrometric analyses of the evolved gases from their cells showed amounts of ^{4}He that were commensurate with their claimed heat. This led Walling and Simons to construct what I've referred to earlier in Chapter III as a three miracle 'non-theory'. In addition to Walling and Simons, Peter Hagelstein of MIT also proposed helium-4 fusion as the source of heat. In response to a question about Hagelstein's work, Simons was quoted as saying "It sounds from what you're telling me like he's (Hagelstein) as smart as we are" (Salt Lake Tribune, April 16, 1989). These proposals of D+D fusion heat without copious amounts of 23.8–MeV gamma rays, neutrons and tritium rank with the proposals of the alchemists who with a mixture of science and magic attempted to change ordinary metals into precious metals.

The rare branch of the D+D reaction proceeds through capture, producing helium with mass number 4 (^{4}He) and 23.8–MeV gamma rays. The above theoretical proposals invoked mechanisms where the reaction energy went entirely into lattice heat, rather than into a photon. Analogies were made with the internal conversion process, and with the Mossbauer effect. Neither of these analogies, however, is applicable in the production of ^{4}He by D+D fusion. In helium the atomic electrons are loosely bound and there cannot be any appreciable coupling between the 23.8–MeV gamma ray and the atomic electrons; therefore, internal conversion or any related process cannot take place at anywhere near the rate that would be required [Fowler, Nature 339 345 (1989)]. Even if internal conversion occurred, the local Utah physicists pointed out that the resulting high-energy electrons would travel several inches into the surrounding water bath and produce a bluish glow known as Cerenkov radiation. Asked whether they had observed this telltale bluish glow, Pons remarked "To tell you the truth, we haven't even looked for it . . . [W]e've turned off the lights in the laboratory and haven't seen anything, but it's not that dark" (*Salt Lake Tribune*, April 23, 1989).

By the time of the Electrochemical Society meeting in Los Angeles (May 8, 1989), the Utah chemists were admitting that their earlier claims of ^{4}He detection were incorrect, citing shortcomings in their instrumentation (ordinary air contains approximately 5×10^{-4}% helium by volume and is a contaminant whenever air is present). This retraction did not, however, settle the helium controversy, because it was postulated that the earlier analyses of the cell gases were based on the incorrect assumption that the helium, if formed by fusion, would come out of the palladium cathodes. At Los Angeles, scientists from other leading laboratories were requesting samples of Utah's palladium cathodes for helium analyses. The same had happened previously at the Dallas meeting of the American Chemical Society. As described in Chapter VI, Fleischmann and Pons rebuffed all those seeking to resolve the question of whether their cathodes contained any helium.

The proof of trace amounts of fusion-produced helium in palladium is made difficult by the fact that helium impurities come from several sources including the air. The necessity of eliminating helium contamination from the air was already shown by the experiments of Paneth and Peters over sixty years ago. If helium, however, was the principal fusion product as postulated by the proponents described above, it should have been easy to establish its presence. Several searches for helium by sensitive mass spectroscopy had already yielded negative results (Lewis *et al.*, Williams *et al.*, and Lawrence Livermore National Laboratory of cathodes supplied by Appleby *et al.*). If the dominant fusion reactions were taking place in the electrolytic cells, then the production of helium and tritium *in situ* in the palladium rods at rates commensurate with heat production would have left a signature easily identified by materials analysis. The Livermore group, using sensitive analysis equipment and noble-gas mass spectroscopy, found neither helium nor tritium above detection limits in a palladium rod, supplied by the Appleby group, from which it was claimed excess heat was generated. Skeptics, including many chemists and most physicists, demanded a mechanism for the Fleischmann and Pons' excess heat. Fleischmann and Pons had invented a black box, supplying D_2O and a small amount of electricity, and extracting large amounts of heat from an unknown nuclear process, presumably deuteron fusion producing helium-4. Enormous pressure was mounting to have the Fleischmann-Pons palladium cathodes analyzed for helium. Having declined all offers of scientists in the United States to conduct the analysis, the Utah chemists in mid-May gave their palladium rods to Johnson-Matthey, a precious-metals company headquartered in London, for helium analysis. University of Utah Vice President Brophy said that Fleischmann and Pons had signed an agreement with Johnson-Matthey to supply palladium cathodes on the condition that they would be returned eventually to Johnson-Matthey for analysis (*Salt Lake Tribune*, May 13, 1989).

Weeks passed without any results from Johnson-Matthey on the helium analysis. Although several American laboratories had said they could do the critical analysis in a matter of days, no results were forthcoming from the British laboratory. My telephone calls to Brophy produced only promises but no results. At the First Annual Conference on Cold Fusion in Salt Lake City in late March, 1990, D.T. Thompson of Johnson-Matthey gave a long technical talk on the autopsy of the Utah palladium cathodes, but gave no results on the helium analyses. Since the Universisty of Utah chemists had claimed watts of excess heat for hours, easily measurable amounts of helium should be observable if the unknown nuclear reaction was deuteron fusion producing helium-4. University of Utah President, Chase Peterson, agreed with Fleischmann and Pons' decision to withhold their information on helium. Peterson said that the University "has to balance the importance of patents and their necessary secrecy against the incredible demand for information that follows such an important break-through . . . [T]he University's loudest critics have come from large, well-financed schools that have sizeable stakes in the more conventional, and more expensive, hot-fusion projects" (*Salt Lake City Tribune*, May 12, 1989).

Responding to increased pressure from the scientific community, the University of Utah scientists agreed on June 12, 1989 to have their used palladium cathodes analyzed by several independent laboratories. The core of the agreement was a double-blind, round-robin assay of the cathodes for helium by six independent laboratories. Dr. John Morrey of Pacific Northwest Laboratory (PNL) was designated to oversee this project and he went to the University of Utah to set up the protocol for the helium tests. Included in the tests were five palladium cathodes, each two-tenths cm in diameter and ten-cm long. Morrey took the samples back to (PNL) in the state of Washington and supervised the distribution of individual pieces of the cathodes himself. The consortium of six laboratories chosen[16] for the helium analysis were:

> Lawrence Livermore National Laboratory
> Rockwell International, Energy Technology Engineering Center
> University of California, Santa Barbara
> Delft University of Technology
> Woods Hole Oceanographic Institution
> Rockwell International, Rocketdyne Division

---

[16] The original plan was to have eight laboratories participate in the study, however, two withdrew after the rods were sectioned.

Each of the five cathodes was divided into ten equal pieces. Pieces of each cathode were sent to each of the above six laboratories and the remaining pieces were stored.

Morrey was playing the very important role as overseer of the project and as the results for the five, as yet unidentified, cathodes were sent to him, he was to collate the data from the six laboratories. After the results were in, the plan was for Morrey to meet Pons at a neutral location where simultaneously Pons would give Morrey the identification and history of each cathode and Morrey would give Pons the results of the helium analysis of each blind palladium cathode.

In response to my many telephone calls to Hugo Rossi, Interim Director of the National Cold Fusion Institute, for information about the helium tests, he wrote to me on September 1, 1989 that he expected the tests to be completed near the end of September. He assured me that he had no objection to my having the results for our panel, and suggested that I make the appropriate arrangements with Morrey to obtain the results of the double-blind helium analyses. After several calls to Morrey in which he wouldn't release any information, I made another call on Friday, October 27, my last opportunity for information before the panel's final meeting in Washington, DC on October 30 and 31, 1989. During this call Morrey revealed to me the following: (a) the analytical results of helium from all six laboratories were in his possession; (b) Pons had received the results on October 6; (c) Pons had not supplied all the necessary history and critical data about each of the palladium cathodes and this had delayed Morrey from mailing the results to the six laboratories; (d) as of October 27, Pons had still *not* supplied all of the required information and Morrey *today* had gone ahead with sending out the helium data to the six laboratories involved in the analyses and (e) he was not free to release the helium data to the general public until he had received the additional information from Pons.

Morrey's decision to mail his collated data on October 27 to each of the six laboratories without all the cathode histories, displayed that he was upset with Pons failure to uphold his part of the double-blind helium tests. I later learned that on October 6, Morrey had met Pons at the Spokane airport to exchange results of the tests and the cathode histories. After giving Pons the experimental results of the six laboratories, Morrey found to his great surprise that Pons had not given him all of the information on the cathodes that had earlier been agreed to in setting up the double-blind experiment.

Dr. N. Hoffman of Rockwell International, Energy Technology-Engineering Center, made some carefully worded statements during one of the discussion periods at the First Annual Conference on Cold Fusion hinting that no helium was found in the double-blind helium tests. Fleischmann jumped up, and to the surprise of the audience, exclaimed that these tests were flawed and should be discarded. I learned shortly later from Morrey that a paper had been written on the helium results and copies were distributed to the principal investigators at the six laboratories as well as to Pons, Fleischmann, Rossi and Brophy at the University of Utah. At this point the law entered the sacred halls of science. Morrey and the other principal investigators received letters from C. Gary Triggs of Morganton, N.C., a childhood friend and lawyer of Pons, declaring that an innocent mistake had been made, and it would not be proper to publish the results - and if they did, they would face legal action. These hardball tactics worked very well; the results of the double-blind helium tests were delayed for many months before being released to the scientific community. In addition, Pons' lawyer, Triggs, accused Morrey of leaking the results of the tests to our DOE/ERAB panel. I can testify categorically that Morrey did not give me any information on the helium content in the cathodes. He kept faithfully his part of the agreement. The same cannot be said, however, for others. On checking with Cheves Walling in April, 1990, I learned that he and others at the University of Utah had read the controversial helium paper, which I and the scientific community at large had been forbidden to see for many months.

In late 1990, the information about the University of Utah's palladium rods supplied by Johnson-Matthey and used in the double-blind experiment was finally released to the public. The history of each of the palladium rods before being sent to the consortium of six laboratories was as follows (*Fusion Technology* **18** 659 1990).

Palladium rod one – ion-implanted with 3 x 10^{-7} mole of ^{4}He at 500–keV energy, then electrolyzed at 800 mA in 0.1 molar LiOD for twenty-eight days. The mean range of 500–keV helium ions in palladium is approximately 0.8 micrometers.

Palladium rod two – original palladium wire as received from the manufacturer, Johnson and Matthey.

Palladium rod three – ion-implanted with 3 x 10^{-7} mole of ^{4}He at 500–keV energy and otherwise untreated.

Palladium rod four – ion-implanted with 3 x 10^{-7} mole of ^{4}He at 500–keV energy, then electrolyzed at 800 mA in 0.1 molar LiOH for twenty-eight days.

Palladium rod five – electrolyzed at 800 mA in 0.1 molar LiOD for twenty eight days.

On November 7, 1989, a month after the October 6, 1989 exchange of information between Morrey and Pons, Pons finally reported that palladium rod five had generated on the order of 5 to 8 milliwatts for twenty-four days. The fact that Pons held back his information on the excess power produced in rod five, the single cathode critical to the test, for one month beyond the October 6 date at which time all the data were supposed to have been exchanged, made the double-blind aspect of the experiment meaningless.

The results of the six laboratories for the three ion-implanted palladium rods showed that the helium content was highly variable from rod to rod. In addition, the subsamples from each of these rods which were analyzed within individual laboratories showed large variations. The helium levels in these ion-implanted rods were so high that laboratories refused to use their best low-level facilities for fear of making these facilities inoperative for future low-level analysis. Fortunately, Morrey had alerted the laboratories to the possibility that some of the rods might contain high levels of helium.

The level of helium in the ion-implanted rods was approximately that expected based on the University of Utah's previous claims of excess heat. However, after Pons and Fleischmann learned on October 6 that the helium level in rod five was at least three orders of magnitude less than expected, they reported to Morrey on November 7, that the heat generated in rod five was milliwatts instead of watts. One can only speculate why this crucial information was not given to Morrey on October 6! The actual level of helium in rod five was of the same order of magnitude as the background sample, rod two. Given the initial helium contamination in the virgin palladium rod two, the variability of the helium implantation procedure, and the failure of the cell (with palladium rod five as cathode) provided by the University of Utah to produce any substantial amount of heat, the only conclusion one can reach is that the double-blind experiment was a colossal failure, known to fail from the beginning and wasting days of valuable research time for six laboratories.

The whole helium-4 (^{4}He) fiasco can be safely summarized with considerable confidence; no helium was found in the cathode of the active cell above the background level. This conclusion followed also from several searches for the 23.8–MeV gamma ray from the $D+D \rightarrow\ ^4He+\gamma$ (23.8 MeV) reaction. All these searches were negative including measurements made in Pons own laboratory by a University of Utah physics group headed by Mike Salamon [Nature 344 401 (1990)]. Furthermore, if one does not believe in miracles two and three discussed previously and accepts the well-known branching ratio for the reaction $D+D \rightarrow\ ^4He$ + gamma ray, the expected helium levels at high power levels is still considerably below the helium contamination of rod two. Knowing the level of helium contamination in rod two, one would have predicted no detectable helium above background even believing miracle one, which was required to produce the claimed high power levels of Fleischmann and Pons.

Most scientists expected that the negative results on helium from the University of Utah's electrolytic cells would end for all time claims for "excess heat" from cold fusion based on the observation of commensurate amounts of ^{4}He. This, however, was not to be the case. In March, 1991 B.F. Bush et al. [J. of Electroanalytical Chemistry **304** 271 (1991)] repeated the early University of Utah claim for ^{4}He (from the D+D reaction) in the effluent gases from electrolysis reactions at palladium cathodes. The editor of the Journal of Electroanalytical Chemistry thought the new claim was so exciting and important that he accepted the authors manuscript without peer review. It is inconceivable that such an editorial decision could be made at this late date in the cold fusion saga. Due to the many ways that one can get spurious results on ^{4}He in the gas phase, one would have expected the authors to have at least obtained confirmatory evidence that they had observed the 23.8–MeV gamma ray from the reaction D+D $\rightarrow$ ^{4}He+gamma ray. This they did not do! Instead the authors chose to repeat the mistake of most cold fusion proponents. After obtaining fragmentary evidence, the authors went to the press before performing the necessary checks and obtaining confirmatory experimental evidence. The validity of the Bush et al. claim requires three miracles, namely the fusion-rate, branching-ratio and concealed-nuclear-products miracles!

Before leaving the subject of fusion-produced helium at room temperature, some words need to be said about the rare isotope of helium, namely helium-3 or ^{3}He. As illustrated by reaction (1) on p. 6, helium-3 and neutrons are among the main fusion products expected from D+D fusion. Insofar that the sensitivity for detection of neutrons is favorable relative to other fusion products, searches for the D+D $\rightarrow$ ^{3}He+n fusion branch have usually utilized neutron detection. Hence, I will not comment further on the use of ^{3}He as a fusion product of importance in searches for D+D cold fusion. What I do wish to comment on, however, is the utilization of ^{3}He measurements for the establishment of upper limits for p+D fusion via the reaction p+D $\rightarrow$ ^{3}He + gamma ray (5.5 MeV). It is known from calculated cold fusion rates of isotopic hydrogen molecules by Koonin and Nauenberg [Nature **339** 690 (1989)] that the rate of p+D fusion is faster than the rate of D+D fusion by some 8 orders of magnitude (10^8). It is not surprising, therefore, that suggestions have been made, that if cold fusion were to occur, it is more likely due to the neutron-free p+D reaction than the D+D reaction, even for cells with small admixtures of H_2O in the D_2O. What is surprising, however, is that such an hypothesis by Professor Julian Schwinger of UCLA was actually submitted and accepted for publication [Z. Naturforsch. **45a** 756 (1990)].

The Schwinger article hypothesized that: (a) the Fleischmann-Pons claim to have produced cold fusion was valid (b) the cold fusion process was powered by the p+D reaction rather than the D+D reaction and feeds on the small amount of H_2O in the D_2O and (c) The p+D reaction does not have an accompanying gamma ray; the 5.5 MeV of energy is directly absorbed by the lattice. Schwinger has elaborated in other papers [Z. Physik D

15 221 (1990)] on the lattice coupling mechanism by which the 5.5–MeV gamma rays are hidden. It is easy to postulate that the p+D fusion reaction is responsible for cold fusion. The hard question is, however, where is the ^{3}He? All searches for ^{3}He have yielded only negative results. For example, helium (^{3}He) analyses of cathodes claimed to have produced forty milliwatts of excess heat showed no ^{3}He to the level of 3.10^5 atoms (September 8, 1989 letter to the ERAB panel from Dr. J.F. Holzrichter, Lawrence Livermore National Laboratory). This is over 10 orders of magnitude (10^{10}) below the level required for forty milliwatts of fusion power for 100 hours as claimed. The six laboratories involved in the double blind experiment supervised by John Morrey also analyzed the palladium rods for ^{3}He. None of these laboratories detected any ^{3}He. If rod five were producing 5 to 8 milliwatts of power for 24.3 days as Pons reported, and due to either the D+D or p+D reaction, an amount of ^{3}He would have been produced well in excess of a million times the experimental limit. On this basis alone one concludes that no nuclear fusion occurred in the Fleischmann and Pons cell.

If the p+D reaction were occurring, the 5.5–MeV gamma rays would also be easily detectable. As explained earlier for the 23.8–MeV gamma rays from the D+D reaction, it is not possible to eliminate the 5.5–MeV gamma rays from the p+D reaction as proposed by Schwinger in (c) above. There are numerous reactions analogous to the p+D fusion process in which gamma rays of comparable energy are emitted, for example, gamma rays from thermal-neutron capture in solid materials. These have been studied thoroughly and there is no evidence for anomalous processes in which a gamma ray of such an energy is totally suppressed in favor of direct conversion into lattice heat. All searches for 5.5–MeV gamma rays have been negative. With both the searches for ^{3}He and 5.5–MeV gamma rays being negative, one concludes that there is no evidence for the reaction p+D $\rightarrow$ ^{3}He + gamma ray (5.5 MeV) being responsible for the cold fusion phenomenon, as hypothesized by Schwinger in postulate (b).

The absence of ^{3}He also placed very low limits on the tritium level in rod five. Since several weeks had elapsed between the electrolysis and the helium analysis, significant amounts of ^{3}He would have formed from the tritium, if present.

C. Neutrons

When the two University of Utah chemists B. Stanley Pons and Martin Fleischmann announced on March 23, 1989 that they had produced nuclear fusion at room temperature, one of their claimed pieces of evidence was that neutrons emanated from their test tube during the electrolysis of D_2O. Neutrons of 2.45 MeV are one of the well-established signals characterizing the D+D fusion reaction (see reaction (1) on p. 6), and if proven to be present, would indeed be evidence for D+D fusion. As described in Chapter VI, however, Fleischmann and Pons' gamma-ray data in support of neutrons were shown by Petrasso et al. [Nature 339 103 (1989)] to be due to instrumental artifacts. In responding to Petrasso et al., Fleischmann et al. [Nature 339 667 (1989)] showed a gamma-ray spectrum that included their assigned signal peak[17] at 2.496 MeV. Such a signal peak agreed with their previously published gamma-ray energy at 2.5 MeV as given by their erroneous equation (vii) [J. Electroanalytical Chemistry 261 301 (1989)]. Hence, it appears that Fleischmann and Pons expected a gamma-ray signal peak to correspond with the neutron energy of 2.45 MeV. However, the gamma-ray peak in Figure 1 of this paper is at an energy of 2.21 MeV in disagreement with equation (vii). It is very difficult to understand Fleischmann's response to Petrasso because by this time he and his collaborators had already written an errata [J. Electroanalytical Chemistry 263 187 (1989)] showing they were aware of the correct value of the gamma-ray energy [see equation (vii) in the errata] associated with the reaction $^1H + n$ (thermal) $\rightarrow D + \gamma$ (although the energy of the gamma ray in the figure of the errata was still at 2.21 MeV). In addition, the shape and intensity of the gamma-ray spectrum (Figure 1) in the errata were markedly changed from those in their original paper. All of these changes in Fleischmann and Pons' claimed signal peak raise serious questions about their treatment of their data. The reply by Petrasso et al. [Nature 339 667 (1989)] was devastating and convincingly demonstrated that the University of Utah chemists had no persuasive evidence that they had actually detected any neutrons. Fleischmann admitted publicly at the Los Angeles Electrochemical Society meeting that the University of Utah neutron measurements were flawed casting doubt on the published claim of producing neutrons during electrolysis of D2O. The question remains about the way some of their early gamma-ray spectra were changed in shape and intensity as well as shifted in energy. Were these changes made to

---

[17] Fleischmann and Pons' letter in Nature and equation (vii) and Figure 1 in their Electroanalytical Chemistry paper and errata indicate great confusion and misunderstanding about the proper energy of the expected signal peak. The technique that Fleischmann and Pons were supposedly using for detecting neutrons from the reaction $D + D \rightarrow {}^3He + n$ (2.45 MeV) was an indirect one utilizing the secondary reaction $^1H + n$ (thermal) $\rightarrow D +$ gamma ray (2.224 MeV), where the characteristic gamma ray resulting from thermal neutron capture on hydrogen would serve as the signal for neutron production. Some of the neutrons postulated to result from the $D + D \rightarrow {}^3He + n$ reaction would be thermalized in the constant temperature water (H2O) bath surrounding the electrolysis cell and subsequently a fraction of these thermalized neutrons would be captured by the hydrogen in water to give gamma rays of 2.224 MeV. Hence, a gamma-ray line at 2.224 MeV is evidence for neutrons.

correct for calibration errors or were they a deliberate move at a later time to place the gamma-ray energy at the known theoretical value for thermal neutron capture on hydrogen? If the latter, this would be a violation of the way science should be done! I tend to believe the Utah chemists were so uninformed about nuclear phenomena and experimental counting techniques, that the whole episode of their mobile gamma-ray peak defining their neutron signal can be written off as due to haste, carelessness, wishful thinking and sloppy science. However, further investigation of Fleischmann and Pons' handling of the mobile gamma-ray signal peak is merited, even though their reported data are due to instrumental artifacts.

Already at the time of the Santa Fe meeting in May, 1989 most scientists agreed that neutrons, if produced at all in cold fusion, are produced at a very low intensity, making the signal difficult to separate from natural backgrounds. Even the proponents of cold fusion who still believe in measurable amounts of excess heat acknowledge that the process is aneutronic. Hence, the excess heat cannot be due to the $D+D \rightarrow {}^3He+n$ reaction. Based on standard nuclear physics arguments, therefore, one concludes that the excess heat cannot be due to the $D+D$ reaction since the neutron branch represents about one-half of the fusion events. A number of believers in cold fusion, however, invoked by fiat a miraculous change in this branching ratio of many orders of magnitude for room temperature fusion in solids. With this contrived justification some continue to associate the excess heat with a nuclear process by claiming helium, tritium or some unknown nuclear process. Since the fragmentary evidence for each of the charged particles postulated to be coming from cold fusion was shown earlier in this chapter to not at all be persuasive, attention is now turned to what evidence exists, if any, for fusion neutrons. This section on neutrons is devoted to an analysis of a variety of experiments searching for the 2.45–MeV neutrons from the reaction $D+D \rightarrow {}^3He+n$ (see reaction (1a) on p. 6).

In many respects neutrons are an ideal fusion product to measure because they have a high probability to escape from the experimental apparatus and, therefore, are in principal easily detected. Extremely low neutron fluxes, however, are difficult to measure because of the large natural background of ambient neutrons and gamma rays. One prolific source of background is cosmic rays, which bombard the earth from space. When the cosmic rays strike any material in the proximity of the experimental apparatus large intensities of secondary neutrons and gamma rays are produced. The intensities of the neutron and gamma ray backgrounds associated with cosmic rays fluctuate with barometric pressure, increasing as the barometric pressure decreases.[18] Due to these fluctuations the background should be

---

[18] As the barometric pressure decreases, there is less air mass to attenuate the cosmic ray flux and the neutron and gamma ray backgrounds increase.

measured simultaneously with the sample and in the same environment in order to replicate the secondary particle fluxes. This has not always been done in cold fusion experiments. Improvement in the signal to background can be achieved by doing experiments deep underground where the cosmic ray flux is attenuated.

The first report of low-level neutrons from room temperature fusion came from Steven Jones and his collaborators at BYU. Their paper was received by *Nature* on March 24, 1989 (a day after the Fleischmann-Pons press conference) and published on April 27, 1989. From the beginning, these researchers claimed no excess heat and made every effort to disassociate their claims of very low-level fusion neutrons from the Fleischmann-Pons claim of large amounts of excess heat from room-temperature fusion. The BYU group detected neutrons in a two-stage neutron counter; first by proton recoil in an organic scintillator, followed within a few tens of microseconds by a signal from the capture of the moderated neutron on boron viewed by the same photomultipliers. This double detection of a single neutron serves to reduce substantially the ambient background due to gamma rays, although there remains background in the experiment due to gamma rays and to real neutrons from cosmic rays and other sources. The measurement of the actual background and the statistical analysis of the data must be done with great care as the rate of cosmic ray neutrons can fluctuate considerably with variations in barometric pressure or with solar activity. These variations and their impact on the Jones *et al.* neutron background were questions raised by our DOE/ERAB panel during its visit to BYU.

The reported evidence for the production of low intensities of neutrons has come from two very different types of cold fusion experiments. The first type of experiment utilized cold fusion cells of the electrochemical type while the second type of experiment employed temperature-cycled high-pressure gas cells. Both types of cells contained palladium or titanium (or both) and large concentrations of deuterium. Stimulated by the early claims of Jones *et al.*, many groups searched for neutrons coming from electro-chemical cells with a palladium (or titanium) cathode and a $D_2O–LiOD$ electrolyte (or some other mixture containing deuterium). As of the end of October, 1989, most experimenters were reporting negative results with upper limits near and below the reported value of Jones *et al.* These data, reproduced from our DOE/ERAB report (DOE/S–0073, November, 1989) are given in Table 2.

TABLE 2. Some cold fusion neutron rates

Authors	Reference	Neutrons per DD pair per sec[a]	Yield Normalized to Jones et al. neutrons[b]
Jones et al.	*Nature* 338 737 (1989)	1×10^{-23c}	1
Mizuno et al.	*J. Electrochem.* 57 747 (1989)	5×10^{-23}	5
Williams et al.	*Nature* 342 375 (1989)		<0.5
Alber et al.	*Z. Phys. A* 339 319 (1989)	$<4 \times 10^{-24}$	<0.4
Broer et al.	*Phys. Rev. C* 40 R1559 (1989)	$<2 \times 10^{-24}$	<0.2
Lewis et al.	*Nature* 340 525 (1989)	$<2 \times 10^{-24}$	<0.2
Butler et al.	*Fusion Technology* 16 388 (1989)		<0.2
Kashy et al.	*Phys. Rev. C* 41 R1 (1989)	$<1 \times 10^{-24}$	<0.1
Gai et al.	*Nature* 340 29 (1989)	$<2 \times 10^{-25}$	<0.02
DeClais et al.	Santa Fe Meeting, May (1989)		<0.01

[a] Assuming that neutrons are produced throughout the volume of Pd or Ti.
[b] For comparison one watt of heat production by D+D fusion would correspond to 0.9×10^{12} in these normalized neutron yield units.
[c] This fusion rate is reported by Jones et al. for run 6. The average fusion rate for all runs is a factor of 6 less.

The results of Gai et al. and DeClais et al., listed in Table 2, are particularly noteworthy because each group sets an upper limit on neutron emission from electrochemical induced fusion that is considerably smaller than the claimed value of Jones et al. Both groups had neutron backgrounds well below that of Jones et al., and raise the question whether the reported neutron signal of Jones et al. may be originating from cosmic rays. The Gai collaboration employed a neutron detector consisting of six large NE213 liquid-scintillator counters with fast photomultipliers. One of the detectors served as the central detector and it was surrounded by the five other detectors. Neutrons and gamma rays were distinguished with state-of-the-art pulse-shape-discriminator electronic modules. The detection of neutrons is characterized as a coincidence event between a primary neutron in the central detector and a scattered neutron detected in one of the five counters in the outer ring. A neutron event is then defined by measuring its pulse shape in the second detector, the pulse heights in the two detectors, and the time of flight between the two detectors, which is related to the energy of the neutron. The coincidence method significantly lowers the background in the detector system while still giving a reasonable neutron detection efficiency. The Gai et al. experiment (a Yale-Brookhaven collaboration) was performed with sophisticated experimental techniques and was the most thorough early search for neutrons from room temperature fusion.

At the time of our DOE/ERAB panel's final report the evidence for bursts of neutrons from temperature cycled high-pressure gas cells was also not persuasive. The initial claim of bursts of neutrons from dynamically induced fusion by the Frascati group [*Europhys. Lett.* **9** (3) 221 (1989)] could not be confirmed in later experiments by this and another Italian group [*Europhys. Lett.* **10** (4) 303 (1989)]. Following the rather sensational claim of the Frascati group, a number of other groups outside Italy also initiated searches for neutron bursts from gas-phase experiments operated under dynamic conditions. A few of these groups reported seeing bursts of neutrons. One such group headed by Howard Menlove at the Los Alamos National Laboratory, in collaboration with the Jones group at BYU, reported positive results already at the Santa Fe meeting. A group at the Sandia National Laboratory, however, working with a more sophisticated detector system having four separate banks of neutron counters, obtained negative results [Butler *et al.*, *J. Fusion Technology* **16** 397, 404 (1989)]. Segmenting the counters into separate banks proved very valuable in assigning spurious events, where only a fraction of the detector bank fired, to background. In view of the difficulty of these experiments in separating the signal from background, our DOE/ERAB panel made the obvious suggestion early on for collaboration between groups reporting positive and negative results. The scientific community was delighted when Jones received an invitation from Gai to do a joint experiment with the Yale neutron detector system.

The new Yale-BNL-BYU collaboration had the potential of solving the question whether neutrons are emitted from titanium in pressurized D_2 gas cells when such cells are temperature cycled. Such a collaborative effort insured uniform cell preparation and experimental procedures that were agreeable both to a group obtaining positive results and a group obtaining negative results. There is evidence that cell preparation is important in the loading of deuterium into titanium by electrolysis [Briand *et al. Phys. Lett.*A **145** 187 (1990)]. These authors reported that a complex mixture of metallic salts in the electrolyte, like in the original experiments of Jones, provokes the formation of electrodeposits which prevent deuterium penetration into the titanium. Joint experiments of skeptics and proponents have a better chance of resolving any such controversial procedural points before the experiment is started. One aim of the Yale-BNL-BYU collaboration was to check again the Menlove *et al.* experimental results. An interim report from Gai *et al.* giving negative results was available on preparing our panel's final report. Shortly after the availability of the Yale-BNL-BYU group's interim report, Jones resigned from the collaboration over differences in the interpretation of the data.

After further analysis of the data in the above interim report, the now Yale-BNL group reported [Rugari et al., Phys. Rev. C **43** 1298 (1991)] no statistically significant deviations from background were observed for correlated neutrons emitted in bursts or for neutrons emitted randomly. Approximately one neutron burst was expected based on the reduced burst rate of Menlove et al. [J. Fusion Energy **9** (4) 495 (1990)] and the relative running times of the two experiments. Anderson and Jones (Proc. of the conference on "The Anomalous Nuclear Effects in Deuterium/Solid Systems", BYU, October 22–24, 1990)] have stated that the Yale-BNL "null" result on neutron bursts does not contradict the results of Menlove et al. (ibid.). There is a clear disagreement, however, in the two groups results on the rates of random neutron emission. The Yale-BNL group's upper limit on the random emission of neutrons was a factor of six to twenty-five times smaller than the range of rates for random emission above background reported by Menlove et al. This significant difference raises the question whether background neutrons are responsible for the positive results of Menlove et al.

A large collaboration of French scientists [D. Aberdam et al., Phys. Rev. Lett. **65** 1196 (1990)] reported no evidence of neutron emission following deuterium loading into palladium and titanium in both electrochemical and pressurized gas-phase experiments. This group reported a very small upper limit to the fusion rate of 2×10^{-26} neutrons per second per pair of deuterons (see Table 2 on p. 141 and Rugari et al., ibid.), a number in the same range as the limit of the Yale-BNL group for random neutron emission.

A workshop on low-level fusion products was held at Brigham Young University on October 22–24, 1990. The proceedings of the Workshop were published by the American Institute of Physics (Conference Proceedings #228 entitled "Anomalous Nuclear Effects in Deuterium/Solid Systems"). Positive claims for low-level neutron bursts and random neutron emission were reviewed, however, there were no new positive neutron results[19] beyond the claims already discussed. The Menlove-Jones group continued to report evidence for neutron bursts at this meeting. The Wolf group, using very large palladium electrodes, concluded that the low-level neutron "effect cannot be considered to be confirmed as yet." Anderson, Johnson et al. reported on an extensive study of the impact of the cosmic ray and other backgrounds on searches for low-level neutrons. In their search for neutron emission from cold fusion systems of both the electrochemical and high-pressure gas cell types, they obtained no evidence for cold fusion processes leading to neutron production. Their paper demonstrated the need for redundant detectors and the exclusion of cosmic-ray backgrounds before any claim can be made for cold fusion neutrons. The

---

[19] Published in leading journals requiring peer-review.

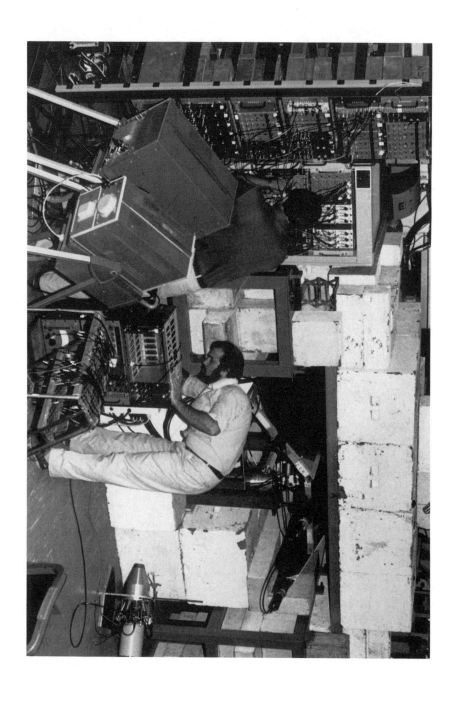

separation of the neutron signal from the background is the major challenge facing those searching for very low-level intensities of neutrons.

Proponents of cold fusion often support their claim by stating that some 100 groups have reported some type of fragmentary evidence for the process. Only a small percentage of these reports have been published in journals with adequate peer review. The fact that not a single experiment in this list has given consistent data (where the yields of fusion products are commensurate with the excess heat) doesn't seem to disturb these proponents. In order to exemplify the nature of the evidence reported by some of the above groups, I discuss first the reported positive results by Jacob Jorne [*Fusion Technology* **19** 371 (1991)]. He reported a fusion rate of 10^{-21} neutrons per D–D pair per second when palladium was exposed to pressurized deuterium gas and the temperature was cycled up to 320 degrees Centigrade (°C).

A few days after Fleischmann and Pons' press conference, Jorne in collaboration with a group of physicists at the University of Rochester embarked on a search for neutrons in an electrochemical type fusion cell. Liquid scintillator (NE–213) neutron detectors and associated electronics were borrowed from our nuclear chemistry group. Dr. Jan Tõke joined the group to insure that the neutron counters were operated properly. This group reported their negative results on cold fusion at the Baltimore APS (1989) meeting. Having established an upper limit for neutrons that was more than 4 orders of magnitude (10^4) below that of Fleischmann and Pons, all members of the research team except Jorne lost interest in cold fusion and returned to their own research activities. Jorne continued to pursue cold fusion leading to the positive results mentioned in the above paragraph. When Jorne obtained his positive results, in late 1989, Tõke immediately checked the neutron counters and found that the thresholds were set too low making the detectors highly sensitive to noise. As can be seen in Fig. 2 of Jorne's later paper, the two identical detectors gave very different results. Hence, these reported positive results should not be taken seriously until they are repeated with properly functioning neutron detectors.

Figure 5. Experimental setup for the detection of neutrons by the Yale-Brookhaven National Laboratory-Brigham Young University collaboration. Shown are part of the massive shielding (some twenty blocks of one ton each), the veto counters, the neutron counters (mostly hidden) and the cells (completely hidden). Professor Moshe Gai (right) and graduate student Steve L. Rugari of Yale University are tuning electronics. Before reporting the group's final results [*Phys. Rev.* **C43** 1298 (1991)], the Brigham Young University members resigned from the collaboration. (Courtesy of Professor Moshe Gai.)

Typical of the cold fusion saga, another rather bizarre story developed from Jorne's experiments. After about one year had elapsed, our nuclear chemistry group asked for the return of our neutron counters for an up-coming experiment scheduled at the GANIL heavy-ion accelerator in France. Eugene Mallove, an ardent believer in cold fusion and the author of a highly positive book [Fire From Ice, John Wiley & Sons, Inc. (1991)] on cold fusion, phoned Jorne and by a series of leading questions implied that I was instrumental in taking the counters away from Jorne and stopping his "successful" cold fusion experiments. On March 1, 1991 Jorne wrote to Mallove to make sure he understood that the neutron counters loaned by the nuclear chemistry group were needed for experiments. Jorne went on to say, "To the best of my knowledge, in no way was this action a result of Professor John Huizenga's stand on the cold fusion issue. In fact, Professor Huizenga was and has remained very informative and helpful regarding the status of my cold fusion activity." Believers in cold fusion, like, for example, Eugene Mallove, are unwilling to accept the negative experimental evidence on cold fusion and look for all kinds of other unrelated reasons to explain why the subject is slowly dying.

As another example of positive results, I discuss the work of a Japanese group. M. Yagi et al. of Tokoku University in Sendai, Japan published two papers [J. Radioanal. Nucl. Chem. Letters 137 411 and 421 (1989)] claiming that small yields of neutrons from the D+D reaction were emitted in titanium (Ti) and silicon dioxide (SiO_2) systems in which D_2 was trapped at approximately atmospheric pressure and studied over a temperature range of liquid nitrogen temperatures (–77°C) to 400°C. W. Meyerhof of Stanford University did a statistical analysis of these data and showed that background counts reported for over fifty different time intervals did not follow the expected Poisson distribution. This is an example of reported positive data on cold fusion that was invalidated when it was examined closely by normal scientific methods.

The above two examples from the cold fusion proponents' long list of claimed positive results illustrate what happens when individual claims are subjected to close scrutiny. It is for this reason that a much more persuasive case for cold fusion could be made if a single consistent and reproducible experiment existed.

There have been many theoretical attempts to "explain" various reported cold fusion results. Most of these "theories" start with the premise that the reported positive findings are true and then proceed to make a series of assumptions in order to "interpret" the claims. Based on a model of palladium metal containing deuterium, Parmenter and Lamb [Proc. Natl. Acad. Sci. USA 86 8614 (1989); 87 3177 8652 (1990)] were one of many groups that calculated fusion rates. The Parmenter-Lamb calculations are noteworthy and mentioned here because Willis E. Lamb, Jr. is a Nobel laureate in physics. In their initial paper these authors deduced a fusion rate in palladium of 10^{-30} neutrons sec^{-1} per deuteron pair, seven powers of ten smaller than the experimental rates reported by Jones *et al.* (see Table 2 on p. 141). In a corrected calculation, these authors obtained a theoretical rate of 10^{-18} sec^{-1}. In a third publication, Parmenter and Lamb further modified their calculation to give a rate of 10^{-23} sec^{-1}, comparable to the reported experimental result of Jones *et al.* These theoretical calculations employ a momentum-dependent effective electron mass. On the basis of their calculations, Parmenter and Lamb "conclude that it would be wrong to say that the [cold] fusion phenomenon is impossible." One problem in comparing the Parmenter-Lamb theoretical fusion rate with that of Jones *et al.* is the fact that the theoretical rate is for palladium metal whereas the experimental result of Jones *et al.* is for titanium metal. Cold fusion proponents have had their claims at any level supported by one or more prominent theorists, including Nobel laureates (see Chapter X for a discussion of the theoretical support of the Fleischmann-Pons phenomenon by Julian Schwinger). This theoretical support is closely entwined with the whole cold fusion saga and has given some credence to experimental claims that are not reproducible and many times lack adequate controls and checks.

In the middle 1980s, Boris V. Deryagin of polywater fame (see Chapter XI) and his collaborators proposed that nuclear reactions occur during the fracture of a deuterium-containing substance. This proposed phenomenon has come to be known as fracto-fusion. It has been suggested that fusion, which would occur at extremely low and undetectable rates under static conditions, might occur at detectable rates under dynamic conditions. For example, if a metal lattice, embrittled by deuteration, were to undergo microfracturing, deuterons might be accelerated in the transient high electric fields across the cracks, reaching energies sufficient to cause fusion. Klyuev *et al.* [*Soviet Tech. Phys. Lett.* **12** 551 (1986)] reported that when a single crystal of LiD was fractured by a device powered by an air gun, a few neutrons appeared to be produced above the cosmic-ray background. Based on these and other results, Price *et al.* [*Nature* **343** 542 (1990)] searched for charged particles from the D+D → T+p reaction during the fracture of LiD crystals. However, they observed no charged particles, casting doubt both on the claim of Klyuev *et al.* to have seen fusion neutrons and on the conjecture that deuterated palladium or titanium could produce a detectable number of particles as a result of fusion at cracks.

Recently, Sobotka and Winter [*Nature* **343** 601 (1990)] repeated an early experiment of Deryagin (*Coll. Jour. USSR* **46** 8 (1986)], in which it was claimed that the fracture of solid D_2O by macroscopic projectiles induced D+D fusion. Sobotka and Winter's results were negative, with an upper limit of the number of fusion neutrons per fracture which was fifteen times less than the positive result reported by Deryagin *et al.* Again, the newer results cast doubt on the claim for fracture-induced fusion.

The above Soviet group also have claimed that fusion occurs in mechanically agitated mixtures of titanium in the presence of deuterium [*Nature* **341** 492 (1989)]. In these experiments titanium chips were agitated in a drum with heavy water and deuterated polypropylene, using steel balls and vibration at 50Hz. Neutrons of very low intensity were reported to have been seen for a few minutes. These experiments need to be repeated by others.

Arzhannikov and Kezerashvili, working in Novosibirsk, USSR, have claimed to see neutrons during the reaction of lithium deuteride (^{6}LiD) and heavy water (D_2O) and during oxidation-reduction reactions of complex salts of palladium and platinum (containing D) reacting with zinc. These authors claim that the small neutron yield from the above reactions disappears when hydrogen (H) is substituted for (D). M. Fowler of Los Alamos has repeated these experiments with null results, obtaining an upper limit on the neutron intensity more than a factor of ten below the Soviet result.

Jones *et al.* in their initial publication [*Nature* **338** 737 (1989)] inferred that products of low-level cold fusion were produced in geologic processes. This is an interesting idea, however, there is presently no persuasive evidence for naturally occurring cold fusion. If it were true, the implications would be major for geophysical problems such as heat-flow modeling, element distribution with depth, and the composition of the Earth's core.

The present evidence in support of fracto-fusion or any of the other chemically related processes described above is not persuasive. Although such processes are sometimes invoked to explain claimed neutron bursts, others doing similar experiments report negative results. For example, Balke *et al.* [*Phys. Rev.* C **42** 30 (1990)] found no evidence for fusion neutrons from D_2, DT and DH pressurized gas cells containing titanium and palladium which were cycled in temperature. Whether there are dynamic conditions where a type of D+D fracto-fusion produces a very low level (but detectable) of fusion products cannot be answered definitely at this time. The experiments giving the reported positive results of low-level neutrons must be repeated by other independent groups. However, based on the many negative results and the marginal statistical significance of the few positive results, I conclude that the experiments to date have not demonstrated a fracto-fusion process.

In searching for very low level cold fusion, it is imperative to perform these experiments with state-of-the-art particle detectors in an environment where the natural background is adequately reduced. In addition, it has been demonstrated that redundant detectors are necessary to eliminate spurious effects. Some of the present experiments are being carried out in laboratories deep underground. One such experiment has been running since January, 1991 in Japan. This collaboration of scientists, which includes Menlove and Jones of the United States, is utilizing the Kamiokande low-background detector in Japan. This detector is a 4500-ton water-Cerenkov system used for neutrino studies. If random or burst neutrons are produced in fusion cells placed inside this detector, they are detected as follows: The neutrons are thermalized and captured by sodium chloride in the water container surrounding the fusion cell producing gamma rays. The gamma rays create electrons in a very large tank of water which give Cerenkov light registered by banks of photomultipliers.

The analysis of the data from the Kamiokande experiment is still in progress. The preliminary results are suggestive of a few low-multiplicity neutron bursts, however, the origin of these events is in no way definitive at this time. Furthermore, since the neutron bursts in the Kamiokande experiment have low multiplicity, they are inconsistent with the very high-multiplicity neutron bursts reported earlier by Menlove et al. [J. Fusion Energy, 9 (4) 495 (1990)].

Although more than one and a half years have elapsed since our DOE/ERAB panel report, the status of the experimental evidence for fusion-product neutrons (either random or bursts) has not substantially changed. As of November, 1989 our DOE/ERAB panel concluded that the experimental evidence for fusion-product neutrons from a new nuclear process termed cold fusion was not persuasive. At the present time there are still no experiments giving consistent and reproducible yields of neutrons above background. This is true despite marked improvements in the experimental equipment and backgrounds utilized in experiments over the last eighteen months. The results of all experiments attempting to find low levels of fusion neutrons remain sporadic with neutron intensities near the background level. One troubling feature is that as experiments become more and more refined, the magnitude and frequency of the neutron bursts diminish. At the time of this writing, a new room-temperature nuclear fusion process producing trace amounts of fusion products has not been established.

IX

Promotion of Claims for Cold Fusion

At a time when the hope of a new source of cheap, clean, and abundant energy was rapidly fading into the sunset, the cold-fusion enthusiasts decided to sponsor a meeting of their own to rally support for cold fusion. The timing of this meeting was well planned to take place on October 16–18, 1989, shortly before our DOE/ERAB Panel was to hold its final meeting in Washington, D.C. on October 30 and 31 for the purpose of completing our final report due by the middle of November, 1989.

The National Science Foundation (NSF) and the Electric Power Research Institute (EPRI) sponsored the meeting which was held in Washington, D.C. at the NSF (the final half day involving the press was held in a Washington hotel). The NSF was established by the federal government with the directive to fund basic research and education at non-profit institutions. Founded after World War II, the NSF is one of the most important government agencies supporting non-mission-oriented research. EPRI is funded by the electric utility companies for the purpose of evaluating and dispersing funds to mission-oriented research programs of interest to the power industry.

This unusual conference sponsored by the NSF and EPRI will go down in history as a striking example of the cold fusion frenzy. Data were presented that didn't have the slightest chance of being correct, defying well-known experimental and theoretical nuclear physics. All of the normal standards and criteria of the scientific process seemed to be expendable when challenged with new 'evidence' supporting cold fusion.

The NSF–EPRI meeting was coordinated by Dr. Tom Schneider of EPRI and Dr. Paul Werbos, Program director for Emerging Technologies in NSF's Division of Electrical and Communication Systems. I mention the NSF Division sponsoring the meeting because other divisions responsible for more basic science at NSF, e.g. the chemistry and physics divisions, wanted to have nothing to do with the workshop and were quite vocal in their opposition. Dr. Marcel Bardon, director of NSF's physics division, sent an electronic mail message to 500 NSF employees stating that "It seems unfortunate that an NSF office is now apearing to encourage such discredited work" (*The Scientist*, November 13, 1989). Dr. Karl Erb, NSF's program director for nuclear physics was quoted as saying that his department was convinced that "there was no reproducible evidence of nuclear physics being involved" in the phenomenon attributed to cold fusion and that he told that to the organizers of the workshop [*Science* **246** 879 (1989)].

The organizers chose to give the NSF/EPRI workshop the title "Anomalous effects in deuterated metals", however, this title did not disguise the fact that the meeting served principally as a platform for those with 'positive' results on cold fusion. The use of camouflage is not unknown when describing research on the periphery of established science. The laboratory at Princeton University, known to the public as PEAR (Princeton Engineering Anomalies Research Laboratory) provides a convenient disguise for activities that are often referred to in emotionally charged terms like "Psychic Phenomena" (*New York Times Magazine*, November 26, 1989, page 50).

The Co-Chairmen of the workshop were Dr. John Appleby of Texas A&M University and Dr. Paul Chu of the University of Houston. Appleby had been a long-time avid proponent of cold fusion. Chu, a skeptic of cold fusion and highly respected researcher for his pioneering work on high-temperature superconductivity, added a degree of authenticity to the meeting. Convincing Chu to serve in this administrative role was an astute move on the part of the organizers. In their letter of invitation to selected participants, the Co-Chairmen wrote:

> This workshop will attempt to achieve scientific dialogue among a small number of invited participants (less than thirty-five) away from the press and other forms of public attention. It is planned that there will be a volume of proceedings which will include recommendations and suggestions for future work. A preliminary agenda, including the names of some invited speakers is being sent to you. The organizers look forward to your input and advice in discussion sessions and in proposing working group reports.

The names of the invited speakers listed in the preliminary program included Fleischmann, Pons, Wadsworth, Oriani, Schoessow, Bockris, Appleby, Huggins, Jones, Wolf, Storms, Menlove, Bard and Teller. All but the last two proposed speakers had previously reported 'positive' results, and Teller had commented positively to the press immediately following the March 23, 1989 Utah press conference. Since participation in the Workshop was by invitation only, at least two skeptics, on receiving the above letter of invitation, refused to attend because, based on the number of "yeah-sayers" on the proposed agenda, they concluded that the meeting simply offered a platform to present again the often-repeated questionable and well-known 'evidence' for cold fusion. Some of the few skeptics who did agree to participate called the organizers beforehand to insure that they were not going to be the lone dissenting voice. On the other hand it was reported that a "believer" had to be persuaded that the meeting was not stacked with skeptics before accepting his invitation.

The charges to the invited conferees were given by Dr. Frank Huband, director of NSF's Division of Electrical and Communications Systems, along with the coordinators Werbos (NSF) and Schneider (EPRI). One of the more unorthodox orders of the organizers forbade the participants to tell the press what they had learned. Such a gag order at a NSF supported scientific meeting evoked the following statement from Dr. Robert Park, the executive director of the office of public affairs of the American Physical Society, "The entire cold fusion episode has been played out against a backdrop of academic misconduct." The skeptics, a token minority at the Workshop, were especially disturbed by the gag order when the organizers later scheduled press conferences and gave their own optimistic versions of the status of cold fusion. Huband defended his decision to hold those meetings with the press on grounds that these press conferences were forced on him because it was the only way to keep the press out of the workshop itself.

The final program included several additional speakers not included in the preliminary program. One of these was Dr. Nathan Lewis, a well-known vocal critic of the cold-fusion claims. Except for one or two other skeptics, the remaining additional speakers had either gone public with positive results or were known supporters of cold fusion including Rolison, Talcott, McKubre, Hutchison, Yeager, Miley, Kim, Whaley, Rafelski and Worledge. It takes only simple arithmetic to calculate that there was a strong intent to have the Workshop consist predominantly of people with so-called 'positive' experimental results along with theorists who could explain such 'positive' results. It was reported (*The Scientist*, November 13, 1989, page 1) that NSF's Werbos in justifying the predominance of "believers" at the meeting stated that the organizers did their best to insure a balanced group of participants, however, a lot of people who were 'nonreplicators' when invited showed up as 'replicators'. It would have been interesting to have had Werbos identify even one of the participants who was converted during that short period of time.

The approximately fifty participants, including the many participants from the NSF and EPRI, were exposed to only a single new experimental claim beyond the reported data that had been previously studied by our DOE/ERAB Panel. This new data, presented by Dr. Debra Rolison of the Naval Research Laboratory, would indeed have been strong supporting evidence for cold fusion, if true. The reported data were, however, so revolutionary that it is extremely difficult to understand how anyone in the audience could have taken the reported claim seriously. Using a technique known as secondary ion mass spectroscopy (SIMS), Rolison and collaborators analyzed the outer surface layers of Pd cathodes electrolyzed in heavy water (D_2O), and reported a startling enrichment in the isotope ^{106}Pd and a diminution in the intensity of ^{105}Pd. The result was promoted to be provocative because of the implication that ^{105}Pd was capturing a neutron to enhance ^{106}Pd. Anyone slightly familiar with nuclear physics knew that a neutron capture process would require completely unrealistic neutron fluxes and would have looked for backgrounds of polyatomic species or experimental mistakes to explain the observed mass spectra. True to past cold fusion history, however, the "believers" at the NSF/EPRI Workshop chose to accept the claimed drastic change in the palladium isotope ratios as the newest 'evidence' for cold fusion. This was in no way the end of the isotope enrichment tale!

Following the severe earthquake in the San Francisco-Oakland area during the late afternoon of October 17, Lowell Wood, of Strategic Defense Initiative (SDI) fame, reached Teller by telephone during the late evening to report that Teller's family was safe. Teller relayed to Wood the unmatched cold fusion claims about the palladium isotope anomalies made at the meeting by the Naval Research Laboratory group. Not to be outdone, Wood recalled some lithium isotope data taken in April at the Lawrence Livermore National Laboratory. Wood told Teller that SIMS analyses of palladium cathodes electrolyzed in D_2O–LiOD solutions produced a fivefold enrichment in the isotope ^6Li on the surface of the cathode. The excited Teller returned the next morning and requested that the Workshop organizers give him time to tell about the fantastic lithium isotope data. In order to explain the isotopic anomalies in palladium and lithium, Teller created a new neutral particle that would catalyze cold fusion by a special type of exotic neutron transfer. Teller[20] went so far as to name the undiscovered neutral particle, "the Meshugatron" (meshuga[21] in Hebrew means "crazy"). This was such a far fetched episode that it was highly newsworthy, and I heard it described on our national public radio station as part of the program *All Things Considered*.

The presence of Teller and Chu at the Workshop served the organizers well. By the end of the meeting both were sufficiently convinced to support optimistic press releases. Chu and Appleby stated in a prepared Workshop press release on October 18 that:

[20] Although Teller suggested a new particle with the appropriate properties to explain some of the above far-out claims, he displayed his skeptical attitude by the name he chose for this particle.

[21] Professor Shaul Mukamel has informed me that meshuga appears in the Old Testament in Deuteronomy 28, Verses 28 and 36; Samuel (I) 21, verses 14 and 15; Kings (II) 9, verses 11 and 20; Jeremiah 29, verse 26; Hoseh 9, verse 7; and Zechariah 12, verse 4. In all of these verses meshuga is used as an adjective meaning crazy. For example, in Kings (II) 9, verse 11, the phrase with meshuga may be translated to "why did the crazy person come to you".

New, positive results in excess heat productions and nuclear product generation have been presented and reviewed in a logical, frank, open, and orderly manner. Based on the information that we have, these effects cannot be explained as a result of artifacts, equipment or human errors. However, the predictability and reproducibility of the occurrence of these effects and possible correlations among the various effects, which are common for the accepted established scientific facts, are still lacking. Given the potential significance of the problem, further research is definitely desirable to improve the reproducibility of the effects and to unravel the mystery of the observations.

Teller was impressed enough to issue a written personal statement to the press in which he offered some very highly speculative suggestions as to how the reported isotopic anomalies might be explained. He stated:

Numerous interesting and partially contradictory results on cold fusion are in disagreement with the solidly established nuclear theory of fusion. There is a possibility to reconcile the results with the theory, assuming that the deuterons act as neutron donors with various materials (other deuterons or lithium or palladium) acting as neutron acceptors. The neutron transfer by direct exchange is prohibited by the Gamow penetration factor, but a catalytic transfer of neutrons might be possible. It is conceivable that the catalyst could be an as yet undiscovered neutral particle.

It is proposed that U–235 be tried as a neutron acceptor because of its great energy release and of its characteristic response to neutron absorption. One may also try to replace the deuteron in its role as neutron donor by beryllium nuclei.

It is recommended in recognition of the high class work that yielded surprising results that the effort be supported in order to obtain clarification, whether the results are due to sophisticated difficulties in the experiments or whether a new phenomenon is involved. An example of such a new phenomenon has been proposed above without claiming that this indeed is the explanation of the results.

Cold fusion theorists were always willing to invoke a 'miracle' when the claimed data were in disagreement with "solidly established nuclear theory." Given a set of reported experimental results, regardless of how preposterous they were, some cold fusion theorists always could 'explain' them. In this case it was not necessary to outwit nature with a new hypothetical particle, because the new 'evidence' was due to a series of errors.

In April, 1989 Dr. G. Bryant Hudson suggested to his colleague Dr. Douglas Phinney at Livermore that he examine the surfaces of some palladium cathodes that had been used in electrolytic cold-fusion experiments at Livermore. Employing the SIMS technique and an ion microprobe with negatively-charged oxygen ions to sputter off very thin layers of the Pd cathodes, Phinney measured the depth profiles of several isotopic species on the surface of the cathodes down to a depth of approximately one-half micron (500 nanometers). The species chosen for study by high-resolution mass spectroscopy were ^1H, $D(^2H)$, $^1H_2$, ^6Li, ^7Li, ^{11}B, ^{24}Mg, ^{25}Mg, ^{26}Mg, ^{106}Pd and ^{108}Pd. The experimental procedure was to cycle through the masses of all the species and record the intensity or number of counts at each mass value. Since the natural ^6Li abundance is more than an order of magnitude less than that of ^7Li, Phinney chose a longer integration time of fifty seconds for ^6Li compared to ten seconds for ^7Li.

Phinney completed these experiments promptly and reported his results on April 21, 1989 to Wood, who expressed great interest and came over to Phinney's laboratory late that Saturday night, both to observe the SIMS technique and to obtain a first-hand description of the final experimental data. Phinney was careful to write on the data sheet given to Wood that the integration time for ^6Li was five times longer than that for ^7Li. Furthermore, Phinney emphasized that the isotopic ratios of lithium, magnesium and palladium were all *normal* within experimental uncertainties. The whole matter laid dormant from Phinney's perspective until he received a telephone call on October 23 informing him about Teller's lecture at the NSF/EPRI Workshop. A meeting was arranged for November 2 in Teller's office with Phinney and several other Livermore employees, but without Lowell Wood. On learning that the lithium isotopic ratios were normal and that Wood's story about the factor of five enrichment in ^6Li was unfounded and had a simple explanation associated with the experimental counting times, Teller expressed his disappointment that the isotopic enrichment results he presented at the Workshop were not true.

I personally experienced some part of this odd episode. Dr. Thomas Finn, the executive director of ERAB, on learning about the lithium isotope anomaly at the Workshop tried to reach me and on learning from my secretary that I was in Livermore, attempted to reach me there. My flight to Oakland, scheduled to land a few minutes after the earthquake, was diverted to Reno and it took an extra day for me to reach Livermore. Finn, however, did reach me at the Livermore Laboratory on October 19, told me ecstatically about the isotope anomalies and asked me to contact Wood. I attempted to calm Finn by explaining that such isotopic enrichments were extremely unlikely and that the reported results were very likely due to a serious mistake having been made by someone. My work schedule did not permit me to contact Wood during my stay at Livermore, however, I asked Dr. C. Gatrousis to check with the experimentalist who had made the lithium isotope measurements and to report his findings to me. I presume my request triggered the telephone call to Phinney on October 23, and brought out into the open Wood's serious misrepresentation of Phinney's data. The members of this Livermore group are world leaders in measuring extremely small isotopic anomalies in nature and were innocent bystanders in the whole lithium fiasco. This ends the lithium saga. The miraculous, factor of five, enrichment of ^6Li in cold fusion electrolytic cells was explainable in terms of a neglected scaling factor correction! Hence, the evidence for lithium isotope enrichment has gone the way of Blondlot's N rays as discussed in Chapter XII.

The story associated with the palladium isotope anomaly is not nearly so interesting, because it was simply due to an erroneous interpretation of data where the experimental mass peaks were misidentified. Contributions from polyatomic species of impurities with masses nearly coincident with those of the palladium isotopes caused the misidentification. In spite of the fact that the palladium isotope anomalies had been discredited for over five months, Bockris submitted a paper on March 26, 1990 [*Fusion Technology* **18**, 11 (1990)] in which he discussed, along with other cold-fusion phenomena, the thermal and 14–MeV neutron induced cross sections on palladium isotopes. He used these mistaken isotopic anomalies data to suggest that the cold fusion reaction is a surface or near-surface reaction, and, therefore, to serve as supporting evidence for his model of fusion. Among cold-fusion enthusiasts mistakes and erroneous results usually decay with a very long lifetime.

Reactions to the NSF/EPRI closed meeting varied widely. Critics suggested that the meeting was cleverly timed to reduce the impact of the anticipated negative report to be issued a month later by our DOE/ERAB Panel. The Workshop was orchestrated toward rendering a favorable judgment on the evidence for cold fusion. Nathan Lewis spoke for the skeptics when he said, "Most of the people there were establishing a foundation for funding with the NSF and not a foundation for the science" [*Science* **246** 879 (1989)]. Lewis' evaluation seemed consistent with Werbos' statement that the purpose of the meeting was to decide what research should be done. On funding Werbos stated, "When we [NSF] get recommendations from the meeting participants, then we will look at the possibility of funding." Dr. Peter Bond of the Brookhaven National Laboratory expressed the opinion that the press conference was far too optimistic in its evaluation of the cold fusion claims, and didn't at all factor in the feelings of the few participants who were skeptics. This is not surprising since the format of the Workshop was designed to do exactly that. Bond went on to say, "The people from EPRI are firm believers . . . [T]hey want it to be true and they're going to will it to be true" (*ibid.*). EPRI had been a major funder of cold fusion research and all indications are that they will continue to do so. One can only surmise that one of EPRI's motivations for sponsoring this Workshop was to develop favorable press as justification for their further financial support of cold fusion.

At the other end of the spectrum of opinions, the proponents of cold fusion called the Workshop "the most scientifically productive meeting yet on cold fusion" (*ibid.*). This is not surprising since the format was highly skewed to so-called 'positive' results. In addition, Edward Teller and Paul Chu, two well-known and respected scientists were convinced to acquiesce, at least to some degree, by subscribing to the proponents' position that excess heat production and nuclear product generation were true by supporting such statements to the press. As an example of a participant pleased with the Workshop, Johann Rafelski in a recent paper [*Fusion Technology* **18** 136 1990)] thanked the organizers of the NSF/EPRI workshop for the "timely setting of the meeting at which [his] work was first briefly mentioned and thanked E. Teller for his inspiring remarks about the 'Meshuganon',[22] a hypothetical neutral particle acting as a vehicle to transport neutrons."

In early October 1989, I invited Professor Chu to give his impressions of the NSF/EPRI Workshop to our DOE/ERAB Panel on October 30. After accepting my invitation, he called back on October 23 to renege on his earlier commitment stating a conflict of interest. In addition, he said he as yet didn't know what the conclusions of the NSF/EPRI meeting would be. I reiterated that I wasn't asking him to give a consensus of the meeting but only his impressions. I failed to convince him to report to our panel. Chu told me that a committee would issue a report in about two months and he couldn't speak for the group now. That was twenty months ago and the report is still not available. It is my opinion that Chu didn't realize what he was getting himself into when he agreed to serve as co-chairman of the NSF/EPRI Workshop, and that he would be used by the organizers to lend his name to a position that he did not fully support.

The strongest public outcry against the closed format of the NSF/EPRI Workshop came from James A. Krumhansl, President of the American Physical Society. In a letter to Mary L. Good, Chair of the National Science Board, Krumhansl wrote:

> The officers of the American Physical Society are deeply concerned by press accounts of a closed meeting sponsored by the National Science Foundation, together with the Electric Power Research Institute, on 16–18 October 1989, in Washington, DC.
>
> The meeting in question was on the subject of "Anomalous Effects in Deuterated Materials," and attendance was by invitation only. Moreover, the fifty scientists who were invited to attend reportedly agreed not to disclose what they learned to the press. In oral and written statements given at an NSF press conference following the meeting, however, the organizers claimed that recent research findings justify additional funding of cold fusion studies. Some of the participants in the meeting have told

---

[22] Different spellings of Teller's neutral particle appear in the literature.

us that they feel these statements violated the agreement, and were one-sided and misleading.

The Council of the American Physical Society plans to discuss this recent episode at its meeting on 12 November 1989. The Society has consistently affirmed its support of the unfettered communication of all unclassified scientific ideas and knowledge, and we are heartened that the National Science Board has also taken a strong position by adopting the eloquent "Report of the National Science Board Committee on the Openness of Scientific Communication."

Dr. Krumhansl raised a number of relevant points in the above letter. Of special significance was the fact that Erich Bloch, the Director of the NSF, had issued just a few months earlier a notice to grantee organizations affirming the commitment of the NSF to open meetings. To my knowledge Krumhansl's letter has never been answered.

At the NSF/EPRI meeting, Dr. Gordon Baym was asked to prepare a summary statement of theory. He had co-authored an earlier paper [*Phys. Rev. Lett.* **63** 191 (1989)] in which he reported a D+D fusion rate many orders of magnitude below Jones' reported rate. In his summary Baym stated:

[W]e are searching for new experimental phenomena in an area in which theory must be supported by consistent, systematic data. Any search for 'anomalous phenomena' is, in its early stages, an experimentally, not theoretically, driven field.

[I]t is necessary to stay as close as possible to conventional physics for as long as one can hold out, and only when driven up the wall should theorists invoke new physics.

This is very sound advice coming from a leading theoretical physicist. As is well known, a number of theorists in the area of cold fusion have chosen instead to ignore conventional physics as again exemplified at this meeting. Theoretical mechanisms were advanced for exotic neutron-transfer processes involving hypothetical new particles such as 'meshugatrons' and 'champs'. As so often had occurred previously in cold fusion research, the reported isotopic anomalies were due to mistakes and the invention of catalytic neutron transfer via an exotic new particle was not necessary.

X

Utah Born and Bred

The day after the University of Utah press conference, Utah's Governor Norman H. Bangerter announced that he would convene a special session of the Utah Legislature in order to request five million dollars to insure that the state of Utah reaps the rewards of Fleischmann and Pons' research at the University of Utah.[23] Some legislators thought a special session was not necessary to discuss funding because waiting until the next regular session would have saved the taxpayer a considerable amount of money. Bangerter's view, however, prevailed and a special session was held on April 7, 1989. The governor took a very supportive stand on the Utah "discovery" and paraphrasing scripture stated that "He that doeth nothing is damned . . . and I don't want to be damned." The governor's position was strongly reinforced by Eugene Hansen, Chairman of the university's Board of Regents. He inspired the state legislators by issuing a warning about the Japanese competition with his statement, "Waiting and seeing could mean the discovery of the century will be developed by Mitsubishi" (*The Scientist*, May 1, 1989).

By a large majority of ninety six to three, the Utah legislature passed the Fusion/Energy Technology Act. This legislation promised research funds, provided that cold fusion was scientifically confirmed. In addition, the Act established a state Fusion/Energy Advisory Council, to be appointed by the governor, and charged the University of Utah to initiate a research and development program in cold fusion. One of the more controversial features of the Act was its provision to prevent the leaking of information on cold fusion. This aspect disturbed particularly the newspaper reporters who hotly protested the decision. The question of what constitutes scientific confirmation received little or no discussion at this time. Had the legislators been attuned to discussions by different faculty members on the campus of the University of Utah, they would have anticipated a problem of whose advice or opinion should be accepted as confirmation of cold fusion. Already at

---

[23] This chapter discusses events associated with Fleischmann and Pons' claim of producing cold fusion in a jar at the University of Utah. The much more modest claim of Jones *et al.* of BYU is discussed in the section on neutrons in Chapter VIII.

this time, very different positions on the D+D nuclear fusion branching ratios, for example, were advocated by Utah Professors Sandquist, Walling and Salamon, with only Salamon defending conventional nuclear physics. Salamon, in accepting the near equality of the yields of the reactions D+D $\rightarrow$ ^{3}He+n and D+D $\rightarrow$ T+p, argued that symmetry is the essence of physics, whereas the other two knowingly invoked a miraculous change in the fusion branching ratios. The 'theory' of Walling and Simons has been discussed in Chapter III. Not to be outdone, Carl Jensen, a physics professor at Salt Lake Community College came up with a 'theory' that was equally as absurd as that of Walling and Simons. In order to explain energy and no particles, Jensen concocted a matter-antimatter annihilation reaction (*Salt Lake Tribune*, April 14, 1989). Although the legislature was almost unanimous in passing the Fusion/Energy Technology Act, there were some lonely dissenting voices, worrying about maintaining some semblance of scientific integrity in Utah.

As established by the Fusion/Energy Technology Act, Governor Bangerter appointed the following nine-member Advisory Council that would control the five-million-dollar fund earmarked for fusion research in Utah (*Salt Lake Tribune*, April 15, 1989).

(1)   Raymond L. Hixson (Chairman) – Chief executive officer of Bonneville Pacific Corporation, an independent producer of electric power plants
(2)   Wilford Hansen – Professor of physics and chemistry, Utah State University
(3)   Karen W. Morse – Chemist and Dean of Science, Utah State University
(4)   Clair Coleman – Chairman of University of Utah College of Engineering Industrial Advisory Board (former president of Quasar Corporation)
(5)   Joseph Gubler – Cedar City accountant
(6)   Ernest Mettenet – Former vice chairman for Hercules Incorporated
(7)   Gary Crocker – President of Research Industries
(8)   Mitchell Melich – Salt Lake City attorney
(9)   Randy Moon – State science advisor

The cold fusion mania in the state of Utah brought out in mass the entrepreneurial element as illustrated by the following ad which ran for a week in early April, 1989 in the business opportunities sections of the *Salt Lake Tribune* and the *Deseret News*:

Fusion research, help design the future. Engineers, scientists, designers, artists, technicians, educators, writers, secretaries, share in equity-royalties, foreign language facility a plus; retired and/or self-funded. Work in Utah. Send resume.

All of those connected with the cold fusion promotion in Utah hoped that the passage of the Fusion/Energy Technology Act by the Utah legislature would serve to catalyze federal funds. In Chapter V, I've described the Hearing Before the United States Committee on Science, Space and Technology on April 26, 1989. University of Utah President Chase Peterson, Fleischmann and Pons along with high-powered consultants appealed to this committee for federal funds to supplement the five million dollars appropriated by the state of Utah. As described previously this appeal by the Utah delegation was unsuccessful in attracting federal funds, even though Utah Representative Wayne Owens had concluded that the University of Utah had hit a home run with the Science, Space and Technology Committee.

Considerable attention was given by the state administrators to the legal aspects of cold fusion. From all external signs, patents had the highest priority. Utah attorney general Paul Van Dam in April sought additional outside expertise in their patent work to protect the state and university interests in the development of cold fusion as an energy source. The state retained the Salt Lake City firm of Giauque, Williams, Wilcox and Bendinger to direct the legal and trial aspects of the patent fight and the Houston firm of Arnold, White and Durkee, patent experts, to secure the national and international patent rights. The California firm with attorney Peter Dallinger, a nuclear physicist who filed the first patent applications, was to continue to assist the legal team (*Salt Lake Tribune*, April 28, 1989).

As of May, 1989, the special nine-member Advisory Council was still grappling with the issue of confirmation, and had not released any funds for scientific experiments. The higher priority given patents and legal issues over science and research came through clearly in the initial release of state funds. An amount of $500,000, ten percent of the state's five million dollars, was allocated solely for attorneys fees to secure the patents that the University was seeking. All of these funds were limited strictly to legal work and could not be used for experiments. Although no money for research could be released until the nine-member Advisory Council agreed that the Fleischmann-Pons claimed nuclear fusion was confirmed, no such restriction applied to legal fees.

By the end of June, the Advisory Council still had not released any research money. Outside advice was sought by the Council on the question of confirmation. Professor John Bockris from Texas A&M was the first to meet with the Council. He was a well-known strong advocate of cold fusion, having by this time reported positive results several times himself. One might have thought that the nine-member Advisory Council would have asked one of the many scientists obtaining negative results to testify also. This, however, was not to be. At their July 11 meeting, the Council chose to invite Professor Huggins of Stanford, also a leading cold fusion proponent, as a witness. If the Council wanted confirmatory evidence, they couldn't have chosen two stronger voices for cold fusion. At this meeting, Council member Wilford Hansen brought up the point that the Council had not as yet defined what is meant by confirmation and he suggested that the Council ought to do that. With Morse absent from this meeting and Hansen wanting more time on the confirmation issue, the Council's decision on funding was postponed until their next meeting on July 21, 1989.

The State Fusion/Energy Advisory Council voted unanimously on July 21 to accept the claims of Fleischmann and Pons on cold nuclear fusion as confirmed. It would be very interesting to know the Council's interpretation of scientific confirmation. Had council members ever had any discussion on confirmation as Hansen suggested? At this time, the bulk of the scientists who had attempted to repeat the Fleischmann-Pons experiment had reported no success. In proclaiming their experiment confirmed, the Advisory Council chose to listen only to Bockris and Huggins, rather than to go through a standard peer-review process with testimony also from skeptics. The fusion fever was still sufficiently high in Utah to enable the Advisory Council to get away with such a superficial and partial examination of the evidence available on cold fusion. Once the Council agreed that cold fusion was confirmed, "they authorized the release of state funds to the University of Utah for nuclear fusion research. However, it withheld approval of budget specifics because of concerns over administrative costs" (*Salt Lake Tribune* July 22, 1989). An executive committee of the council had constructed a rather detailed budget which the entire council could not agree on. One of the problems for some council members was the large allocations for facilities, administration and legal fees. Finally in early August the Advisory Council agreed to release the funds, some four months after the Utah Legislature had in principal allocated the research funds for cold fusion. At this stage of negotiations the University of Utah had eliminated the $400,000 appropriation for legal expenses beyond the $500,000 already set aside for attorney's fees (*Salt Lake Tribune*, August 1, 1989). All of the members of the Advisory Council approved the revised budget except the two scientists on the Council from Utah State University, Wilford Hansen and Karen Morse. Hansen abstained in the vote because of a conflict of interest caused by his submission of a proposal for funds to support his own research. Morse voted against approval of the budget because she wanted a larger fraction of the funds spent on research rather than administration and facilities. This seemed to be a legitimate concern because many scientists thought that the fundamental issue was to establish confirmation of cold fusion by a massive research effort. The issue of confirmation raised its ugly head once again when Morse stated "We can only have verification [of cold fusion] when a full understanding of the science is in" (*Chemical and Engineering News*, August 14, 1989). The majority of the Council members, however, were satisfied by Peterson's promise that the administrative costs would not be used for promotion but to lure federal and private funds to the new National Cold Fusion Institute (NCFI) located in a modern building at 390 Wakara Way in the suburban-like University Park. The NCFI was officially established on August 14, 1989. The expectation at this time, as scientists were moving into this recently acquired Institute building with approximately 25,000 square feet of area, was that a sizeable fraction of its budget would come from federal and private sources, increasing substantially the second year.

The final deliberations and forthcoming decisions by Utah State's nine-member Advisory Council on the confirmation of Fleischmann and Pons' cold fusion experiment and the proposed release of Utah's five million dollars for cold fusion research were taking place concurrently with our DOE/ERAB panel's work on our interim report, completed at the July 11–12 meeting and published early in August. It seems clear that the Utah Advisory Council in reaching positive decisions on both confirmation and release of funds was intentionally snubbing the findings of our panel, since their actions were in direct contradiction to our panel's conclusions. I quote from our interim report:

The panel finds that the experiments reported to date do not present convincing evidence that useful sources of energy will result from the phenomena attributed to cold fusion. Indeed, evidence for the discovery of a new nuclear process termed cold fusion is not persuasive. Hence, no special programs to establish cold fusion research centers or special programs to support new efforts to find cold fusion are justified at the present time.

Our DOE/ERAB panel was constituted to give advice to the federal government and, hence, had no direct responsibility for providing advice to Utah. Even so, there was concern for spending prematurely large amounts of Utah State's public monies. Those in authority in Utah didn't want to hear anything negative about cold fusion and, not only did they ignore the DOE/ERAB panel's advice, some were downright hostile to the DOE panel. Utah Senator Jake Garn, for example, had many harsh things to say about our panel, but acknowledged that our published findings had likely closed the door for federal funds supplementing state funds in support of the National Cold Fusion Institute. In this respect Garn had a more realistic view on federal funding for cold fusion research than local administrators who were proposing budgets with sizeable federal and private funds. Garn on referring to criticisms of Fleischmann and Pons' claims remarked (*Salt Lake Tribune*, July 13, 1989):

> I used to think politicians were dirty, conniving, dishonest people. I've decided that scientists are far worse than politicians with what has gone on . . . down the road there's going to be a lot of scientists at MIT and other universities that are going to be very, very embarrassed at some of their comments that they have made.

Such uncritical strong support for cold fusion without confirmatory evidence by many in Utah developed into a paranoia with the villains being the "Eastern Establishment" universities, e.g. Harvard, Yale, MIT, etc. Pons stated publicly that our DOE panel was mandated to come up with a negative conclusion and agreed with Bockris' labeling of our panel as a "killer commission."

Pons was elated by the decisions of Utah's Advisory Council since he interpreted their actions as vindicating him. Making decisions about important scientific claims by democratic vote of committee members who were mostly nonscientists was a risky procedure, especially when the stakes were so high. But to Pons it was a victory because the idea of cold fusion was "Utah born and bred"! It was a surprise, therefore, that Pons and Fleischmann chose not to move their laboratories to the newly created National Cold Fusion Institute, but to stay in the Henry Eyring Chemistry Building. This lack of participation by Fleischmann and Pons in the day to day activities at the Institute caused Utah State legislators considerable concern. Vice president for research, James J. Brophy, defended the two scientists by saying "they're not team research players . . . they are sharing their scientific data with the institute, but only that which they have chosen to share . . . Pons and Fleischmann have the characteristic of working close to their chest" (*Salt Lake Tribune*, October 19, 1989).

Several months elapsed before the National Cold Fusion Institute (NCFI) was able to attract a permanent director. Hugo Rossi, Dean of Science at the University of Utah, served as the first interim director. He conjectured that the recent bad press on cold fusion had been very harmful in their efforts to hire a director and staff for the institute. In summarizing the progress of work at the institute in late September, 1989, Rossi reported that their cells had neither produced excess heat nor fusion products such as tritium and neutrons. As already discussed in Chapter VII, Rossi's statements were similar to those of our DOE panel. In addition, he stated (*Salt Lake Tribune*, September 26, 1989):

> It's gotten to be time for us to start wondering if we're doing anything wrong . . . We have a conference coming up . . . if we don't have any

papers to present, then this place will be closing up shop. I'm not saying I will do that. I'm just saying I think that is what would happen.

These negative, but honest and straightforward statements about the current status of cold fusion experiments at the NCFI were not well received in Utah. This was unfortunate because it appeared to scientists outside Utah that Rossi was attempting to give the NCFI some scientific credibility. As a result of the media's coverage of Rossi's statements, University officials forbade future contact of institute employees with the press without approval. This whole controversy with the press illustrates the tensions caused by promoting a particular scientific position and restricting open communication. Although the University issued no formal announcement, Dr. Rossi resigned in November, 1989 and James J. Brophy, Vice president for research, replaced him as interim director.

Brophy, a physicist, served from the beginning as chief spokesman for cold fusion at the University of Utah. He was the defender and promoter of cold fusion and the buffer between the scientific community and Fleischmann and Pons. When our DOE/ERAB panel visited the laboratories of Fleischmann and Pons, arrangements were made through Brophy. He always portrayed the optimist with child-like faith that cold fusion was a fact of life and the rest of the world would eventually have to come around to his and the University's official point of view. When Rossi told the press that the NCFI had not been able at that time to confirm the Fleischmann-Pons experiment, Brophy told the legislators that even though no statistically significant events were seen, the University officials were encouraged by reports from a number of other laboratories that had reported seeing some aspects of the Fleischmann-Pons experiment. He consistently found something positive to say about cold fusion. When our DOE/ERAB panel requested information about recent experimental results at the NCFI prior to writing our final report, Brophy excused their lack of participation by telling the press "some of the things they were asking for were simply not available . . . they were asking for information in our patent applications which we did not think we could release to them" (*Salt Lake Tribune,* October 31, 1989). That story no doubt played well in Utah, even though it wasn't factual. Our DOE panel was requesting scientific data to establish whether or not hard evidence was available to support cold fusion, and had no interest in anything pertaining to patents. Brophy's unquestioned loyalty to the principal players involved in Utah's cold fusion drama did, however, have its downside. Those making decisions at the University of Utah had lost contact with the scientific community at large, especially the physics community, and Brophy was of no help. As a physicist, he allowed the University of Utah to go public with cold fusion without ever having consulted anyone in the physics department. This was an inexcusable mistake!

Although the official position on cold fusion at the University of Utah was unswervingly positive, there were signs that numerous faculty had broken with the party line. This had been true for the Utah physics faculty from the beginning, but by September many other faculty had joined the ranks of the skeptics and were worried about the reputation of the university. A measure of the contempt toward cold fusion by campus faculty showed itself on September 22, 1989 during the lecture on cold fusion by Douglas Morrison, a CERN particle physicist who had written and distributed a classic series of notes on cold fusion. Morrison, a leading critic of cold fusion, portrayed a very negative picture of the evidence for both fusion products and excess heat in his lecture. In addition, he presented an interesting analysis of the regionalization of cold fusion results by dividing the world into two zones. "Zone A comprises northern Europe and the major United States laboratories and parts of North America where the New York Times is the dominant serious newspaper. Zone B is the rest of the world." Based on this simple division of the world into two parts, Morrison arrived at the following regionalization of negative and positive results.

TABLE 3. Regionalization of negative and positive results
on cold fusion[24]

Date	Zone	Number of Positive Results	Number of Negative Results
As of May 2, 1989	A	1	18
	B	25	2
May 3 – May 24	A	2	16
	B	6	11

Morrison interpreted these figures near the end of his lecture in terms of Pathological Science (discussed in Chapter XII) where the history of an erroneous result goes through different phases. His point was that Zone A scientists had the advantage of both more sophisticated equipment and the luxury of assembling quickly strong interdisciplinary teams of experts, hence, these groups quickly reached the final phase of an avalanche of negative results. On the other hand, Zone B scientists as of May 2 were still

[24] Douglas Morrison, Cold Fusion News No. 20, October 20, 1989.

in the initial phase with almost all confirmatory results. Later in May, Zone B was moving into the intermediate phase with about equal numbers of positive and negative results. It was reported in the *Salt Lake Tribune* (September 26, 1989) that the faculty gave Morrison a standing ovation at the end of his lecture, a most unusual tribute following an academic seminar. This spontaneous reaction gives one an indication of the depth of the negative feeling by campus faculty on cold fusion. Faculty opinion was, however, quite different from that of the general population in Utah. On September 24, 1989 the *Deseret News* published a public opinion poll of people in Utah as to the public perception of cold fusion. The believers, disbelievers and undecided comprised 61%, 15% and 24% percent, respectively. If such a poll had been taken at this time in any other state, my guess is that the percentage of believers would have been much lower.

The Fusion/Energy Advisory Council continued to approve the scientific efforts of the NCFI with its staff of approximately thirty scientists and technicians, some with joint appointments on campus. The institute was operating with no federal funds, no corporate patrons and on a scale much reduced from the grandeur envisioned initially. Only part of the laboratory space was occupied. The equipment comprised mainly electrochemical cells and elementary nuclear counters. There was little in terms of radiation monitoring and facilities for handling high-level radioactivity. The laboratories had the general character of standard chemical facilities with no regard for high level radiation. At least by this time no one had any grandiose illusions that dangerous levels of nuclear fusion products would be generated from the electrolysis of D_2O.

Near the end of 1989, it was announced that Dr. Fritz G. Will had accepted the position of the first director of the NCFI and would take over on February 1, 1990. Dr. Will, a native of Germany, was educated in Germany with a Ph.D. in physical chemistry in 1959 from the Technical University in Munich. Dr. Will had spent some three decades at General Electric in Schenectady, New York. It soon became clear that he adopted the Fleischmann-Pons belief that despite little or no evidence of nuclear processes, the excess heat from electrolysis of D_2O with palladium electrodes was too large to be anything but a nuclear reaction. Will stated that "those kinds of thermal energies are way beyond any chemically, mechanically or metallurgically explainable energies" (*Salt Lake Tribune*, December 20, 1989). The acceptance of the reality of cold nuclear fusion on such flimsy evidence is hard to understand. The thought that the claimed excess heat was due to an erroneous cell calibration or other such error did not seem to be worthy of consideration.

One of the many research groups searching for fusion products in electrolytic cells containing palladium cathodes and a D_2O–LiOD electrolyte was a University of Utah physics group headed by Michael Salamon. This group found no evidence for nuclear fusion, like so many other groups our DOE/ERAB Panel was in contact with. The difference between the Salamon and other groups, however, was that the Salamon group monitored the cells in the Fleischmann-Pons laboratory. Their results were astonishing, no fusion products were observed! During the preparation of our final DOE report, I learned about the results of Salamon and his nine co-authors by telephone conversations and through an early draft of their manuscript which was later submitted to, accepted, and published in *Nature* [**344** 401 (1990)].

Salamon's group was asked by University of Utah officials, and allowed by Pons, to make independent measurements of any radiation emanating from the Fleischmann-Pons operating cells. Using a lead-shielded sodium iodide (NaI) detector (of dimensions eight by four inches) that was immediately available to them, they proceeded to install it directly under the table housing four of the Fleischmann-Pons open cells. This detector was operated in a dynamic range of gamma-ray energy of 0.1 to 25.5 MeV. Gamma-ray data were collected over a five-week period (May 9 to June 16, 1989). In addition, several neutron detectors, made of ^{235}U foils sandwiched between nuclear-track-detecting plastic film, were placed within the water tank adjacent to the cells. Absolute efficiencies of these detectors were determined with standard gamma ray and neutron sources.

Limits were placed on the fluxes of neutrons, protons, gamma rays and electrons that were produced in the Fleischmann-Pons cells over the above five-week period. The reactions, measurement techniques and results of Salamon's experiments were as follows.

(a)   Neutron flux limit from the reaction $D+D \rightarrow {}^3He+n$.

The 2.45–MeV neutrons expected from this reaction were searched for by the 2.224–MeV gamma ray that is produced following the thermalization and capture of the neutrons by the hydrogen in H_2O. These measurements gave an upper limit to the power level of 10^{-11} watt.

(b)   Proton flux limit from the reaction $D+D \rightarrow T+p$.

The 3.02–MeV protons expected from this reaction were searched for by the low-energy gamma rays associated with radiative de-excitation of the even-even palladium isotopes following Coulomb excitation by the 3.02–MeV protons. These measurements gave an upper limit to the power level of 10^{-2} watt. A more stringent limit on this reaction of 10^{-6} watt results by searching for the 14–MeV neutrons from the reaction $T+D$.

(c)   Monoenergetic 23.85–MeV and 5.49–MeV gamma ray flux limits, respectively, from the reactions $D+D \rightarrow {}^4He+\gamma$ and $D+p \rightarrow {}^3He+\gamma$.

A direct search for these high energy gamma rays gave upper limits to the power level from these reactions of 10^{-11} watt.

(d) Internal-conversion electron flux limits from the reaction $D+D \rightarrow$ ^4He+e.

An upper limit to the power level from this reaction was 10^{-8} watt. The absence of a neutron signal with the nuclear track detectors, allowed an even lower limit to the power level of 10^{-12} watt.

Although the Salamon team placed extremely low limits on the power level of the Fleischmann and Pons cells over an early five-week period of the lifetime of cold fusion, the cold fusion controversy was in no way settled. Pons claimed that none of the electrolytic cells in his laboratory were, at the time they were being examined by Salamon's group, producing any excess heat. (This is inconsistent with Pons statement on August 16, 1989 at the EPRI conference at the University of Utah where he stated that there was a low-level heat excess during this period.) There was a two-hour period, however, during which time Pons reported an excursion of excess heat in one of his cells. It was during this very period, that a lightning strike had cut the power to the physicists' counters and made them inoperative. Pons claimed, therefore, that the nuclear products could not be measured during the heat burst. Salamon *et al.* had a clever answer to this most unusual coincidence. During the time that the power was off and the two-hour segment of excessive thermal release from Pons' cell was occurring, Salamon's large sodium iodide crystal under this cell would be neutron activated producing radioactive ^{24}Na. Since ^{24}Na decays with a 15.0-hour half-life, a signal would be present for several days, if ^{24}Na had been produced. Since no signal was observed, Salamon *et al.* were able to set an upper limit on the power level, during Pons' reported heat excursion, of 10^{-6} watt for the reaction $D+D \rightarrow {}^3He+n$. In summary, during the five-week period in May and June, 1989, Pons cold fusion cells which may or may not have been producing excess heat certainly were not producing fusion products as predicted from conventional fusion reactions.

Pons' reaction to the negative results on fusion products obtained by Salamon and his colleagues was predictable based on an earlier similar situation. He first argued that the Salamon group failed to conduct measurements over a long enough period of time, since his own experience had shown that nuclear fusion processes occur in bursts over periods of months. Then he volunteered that a two-hour heat excursion occurred while Salamon's counters were inactivated. Apparently Pons had not anticipated Salamon's shrewd response. All of Pons' attempts to negate the impact of the results of the Salamon group is reminiscent of his handling of the double-blind helium experiments. In this case when he became aware that the active palladium electrode contained no observable helium, he announced that the power level was only a few milliwatts, too low for producing observable helium.

The publication of the paper in *Nature* on March 29, 1990, by Salamon *et al.*, during the time of the First Annual Conference on Cold Fusion, precipitated a number of accusations and actions by Fleischmann and Pons. They said the paper was "factually inaccurate", exhibited "some selection of data", showed "serious inconsistencies" with other experiments, focused on questions "predesigned to provide the negative results reported" and was "engineered" by *Nature* to appear on the date of the First Annual Conference on Cold Fusion (*The Chronicle*, June 6, 1990, page A6). An even more serious manifestation of Fleischmann and Pons' displeasure with the paper, surfaced a few days later when Salamon and eight of his co-authors (including an undergraduate and graduate student) received threatening letters from attorney C. Gary Triggs. Triggs, acting for Fleischmann and Pons, declared that the "paper as published was untenable" and "should be voluntarily retracted." He went on to say that he had been "instructed by his clients to take whatever action is deemed appropriate to protect their legal interests and reputations" [*Nature* **345** 561 (1990)]. Triggs also claimed that *Nature* was trying to sensationalize Salamon's negative results by publishing at the time of the Conference. In actual fact, the paper had been accepted for publication on January 30, 1990 and it was by coincidence that it appeared near the end of March.

Salamon and his co-authors felt that Triggs' letter represented a direct attack on their academic freedom and was antithetical to the spirit of free academic inquiry. Adding insult to injury, they were told that legal representation by the University could not be guaranteed. This was especially a serious mishandling of events by top University of Utah officials, since these scientists had been requested by the administration to make nuclear measurements of the Fleischmann-Pons cells. Eventually, through the efforts of Joseph Taylor, vice-president for academic affairs, the university reversed its position and guaranteed the scientists legal indemnification. Later Tim Fitzpatrick, reporter for the *Salt Lake Tribune*, reported that Triggs, a friend of Pons, had been paid $68,000 by the University, while at the same time sending threatening legal letters to university employees. Eventually the University terminated its financial relationship with Triggs, but did not recover its fees.

The First Annual Conference on Cold Fusion, sponsored by the National Cold Fusion Institute, was held in the University Park Hotel, a few yards away from the NCFI, on March 28–31, 1990. Some statistical data on the conference are given in Table 4.

TABLE 4. Participant and paper distribution at the
First Annual Conference on Cold Fusion

Total participants	296[25]
NCFI	22
Local (Salt Lake City) Participants excluding NCFI Participants	82
All other US participants	167
Foreign Participants	25
Total number of Papers	38
United States Papers	32
India Papers	2
Italy Papers	2
England Paper	1
Taiwan Paper	1

As can be seen from the statistical data in Table 4, the conference was dominated by speakers and participants from the United States. Contributions from Northern Europe were completely missing (the paper from England was from the Johnson Matthey Technology Center and described palladium rods returned from Pons and Bockris). Also missing were papers from the USSR, China and Japan. The lack of papers from Japan was inconsistent with propaganda from Utah and from stories written by selected reporters claiming spectacular advances in cold fusion by the Japanese. For example, the *Wall Street Journal* ran an article on December 1, 1989 entitled "Japan's Cold Fusion Effort Produces Startling Claims of Bursts of Neutrons." Again on March 2, 1990, the *Wall Street Journal* ran another positive article on U.S. results entitled "Cold Fusion Research Dispels some Doubt." The latter article featured work by Storms and Talcott on tritium. As with most of Bishop's pieces in the *Journal,* he took the position of a believer and didn't consult with experts who had not found tritium production in cold fusion. In this respect the *Wall Street Journal's* technical section differed from the *New York Times* in their stories about cold fusion. If the subject was known to be controversial, the *Times* took the trouble to consult experts with other views. One conclusion comes through clearly from the data in Table 4, cold fusion after one year was essentially an United States phenomenon, except for pockets of activity in India and Italy.

[25] This number is from the list of participants as published by the conference organizers. The number attending the sessions was approximately three-quarters of this number.

Participation in the First Annual Conference on cold fusion was an unique experience, one very different from that of any large scientific meeting I've ever attended. A religious-like fervor prevailed in most of the scientific talks, all of which were given by people who were believers. In spite of inconsistencies and direct contradictions between the experimental claims presented and standard fusion theory as well as all pre-cold-fusion experimental results, most of the time by many orders of magnitude, the overwhelming majority of believers adopted the viewpoint that they were dealing with a completely new phenomenon when describing the claimed room temperature nuclear fusion of hydrogen isotopes in solid lattices.

In some respects the conference was a much enlarged version of the NSF/EPRI Workshop. There were, however, major differences, the most important of which was that skeptics or nonbelievers and media people were allowed to attend, although the media people were not allowed to be present in the lecture hall. This restriction was not important since rooms with live television coverage were available. Another difference was that no talks were presented by nonbelievers, whereas there were a few such talks at the NSF/EPRI Workshop. The total number of skeptics and non-believers registered for the conference was very limited, possibly between ten and twenty. This number, however, was not controlled in any way and simply reflected the fact that most of the large number of scientists who had obtained negative results had simply lost interest in cold fusion and were back doing their pre-cold-fusion research.

Dr. Fritz G. Will, the first Director of the National Cold Fusion Institute, opened the conference with a highly positive address repeating the believers' belief that recognized experts in several fields of science had confirmed the Fleischmann and Pons results in many countries. He also repeated the proponents belief that "the multitude of results obtained by so many different groups can no longer be explained away as experimental artifacts." One of the more incredible statements in his opening address was:

> At this conference, another significant set of presentations will occur. Theoretical physicists will present novel theoretical models aimed at explaining why nuclear fusion can occur in solids, where classical nuclear physics (applicable to gases) fails to provide explanations.

The experimental papers, comprising about two-thirds of the presentations, were a series of isolated talks, each reporting fragmentary evidence of cold fusion. After one year there was still no single convincing experiment, carried out with the necessary checks and controls, that gave consistent and reproducible results! Contrary to the above statement in Will's opening address, the proliferation of theoretical proposals to explain cold fusion left the theory in a state of chaos at the Conference, requiring a series of various miracles to 'explain' the claims of excess heat, the paucity of fusion products, the exotic branching ratios and the direct coupling of the reaction energy into the atomic lattice. The first question needing an answer was, who decreed that all aspects of classical nuclear physics was only "applicable to gases"? The believers at the Conference simply ignored knowledge of nuclear physics gained over the last fifty years and declared that this knowledge was not applicable to nuclear reactions in solids!

The believers lack of respect for conventional nuclear physics made interaction with skeptics in the audience during the question periods essentially inoperative. At the start of each session the chair read out the conference rules forbidding audio and video recording (though the organizers made such recordings) and stating that the questions had to be addressed specifically to the subject presented. Steve Kellogg of Caltech and Rich Petrasso of MIT, both members of large research teams that each did an extensive series of experiments on cold fusion, all with negative results, asked penetrating questions that received only superficial and unsatisfactory answers, sometimes cut short by the chair. Will was so disturbed by their questions, which I thought were insightful and highly appropriate, that he prefaced one of the press conferences with a long prepared statement lashing out at them and denouncing their critical questions and appealing to participants to offer constructive suggestions and comments. Dr. Will's attack prompted Hugo Rossi, Will's predecessor, to clarify the purpose of the conference and he defended the questioners with "We welcome their criticism."

The way the organizers manipulated the press conferences was most interesting. There were up to one hundred media people in attendance, most being doubtful, if not outright skeptics of cold fusion. Since the experts to be questioned at each press conference consisted of Will, the session chair and the session speakers, there was a built-in mechanism to insure that the press questioned only believers. At no time did a skeptic (Nate Hoffman said he was neither a skeptic nor a believer) have an opportunity to participate in either of the two special panel discussion sessions or in a press conference. The media expected Fleischmann and Pons to participate in at least one press conference, and expressed their displeasure when told that this would not happen. Following each scientific session, Fleischmann and Pons would quietly slip away to avoid the press who had a strong presence in the corridors outside the lecture hall. A number of the media were even more incensed when they learned that there had been a private press conference for a few selected journalists. This attempted control of the press by NCFI officials backfired at times and caused some eruption of tempers. For example, when Bockris discussed his evidence for tritium production over a consecutive twelve-day period and then proceeded to display a graph with a sharp peak followed by a strongly decreasing curve, one journalist fired back – "you think we are stupid and do not realize that your graph should rise continuously over the whole period if tritium is steadily produced, instead what you have shown us is a high rate for a short time and then a decay!"

One of the journalists attending the First Annual Conference on Cold Fusion was Jerry E. Bishop, Deputy News Editor of the *Wall Street Journal*. A few days prior to the conference, it was announced that he had won the American Institute of Physics (AIP) Science Writing Award for his series of articles on cold fusion over the first year. As soon as this was announced, I immediately contacted Professor W. Peter Trower, a physicist and one of the judges for the AIP Award. In addition, I contacted the Office of Public Affairs of the American Physical Society. I was upset and particularly interested in learning about the criteria that were applied in making the Award. I had read Bishop's series of articles on cold fusion, thanks to one of my neighbors, who, knowing my involvement in the DOE/ERAB panel, had kindly supplied me with the Bishop *Wall Street Journal* clips. Bishop was one of the first journalists reporting on cold fusion. He exploited his contacts in Utah to produce newsy and informative pieces about which one could make many positive statements. From my perspective, however, Bishop's articles painted a far too optimistic picture of cold fusion. They lacked the scientific balance expected of the leading financial paper in the United States. He quoted extensively from the leading believers of cold fusion, always hinting that they were on the threshold of the discovery of the century, while essentially ignoring the vast majority of scientists getting negative results. On October 8, Bishop did devote a couple of sentences to the interim report of our DOE/ERAB panel, however, he buried them in a long article quoting the 'yeah-sayers' without any evaluation of their claims. More importantly, he was completely silent following the issuance of our negative final report, an extensive study of cold fusion phenomena by the only independent national panel investigating the subject.

On March 20, 1990, Dr. Robert L. Park, Executive Director of the Office of Public Affairs of the American Physical Society (APS) wrote a letter to the Manager of AIP declining his invitation to attend the luncheon honoring Bishop for the AIP Science Writing Award. In addition, several other officers of the APS boycotted the ceremony (I also declined my invitation to attend). Park expressed the sentiments of many of us when he wrote:

> I cannot imagine on what basis Bishop was selected. In reviewing his "cold fusion" articles, I note that he was frequently the first to report some new claim, but that is hardly a substitute for accuracy. His information came almost exclusively from the proponents of cold fusion, and I see little evidence that he ever questioned the information he was given, or sought contrary views.

Bishop's reports of extravagant claims were rarely followed up when the evidence came apart. On May 10, 1989 he wrote an article with the lead "Fusion Brouhaha May Be Settled Soon by Helium Test," in which he discussed the decisive test for fusion: is there helium in the Pons and Fleischmann cathodes? But when Pons and Fleischmann refused to reveal the results of the helium assay Bishop remained silent.

I do not know what safeguards the AIP employs to guard against this sort of embarrassing blunder, but they were clearly inadequate. The worst part of it is that it devalues the Award in the future.

In response to the concerns of many of us, the American Institute of Physics (AIP) has changed its rules on the journalism award. In the future the awardee selected by the AIP judges[26] will have to be approved by the AIP Board of Governors.

It was at the First Annual Conference on Cold Fusion that I met and had my first one-on-one conversation with Bishop. I had to restrain my desire to review with Bishop what I thought were serious deficiencies in the impressions created by his pieces. I limited my criticisms, however, to his snubbing of our panel's final report and proceeded to give him a copy and explain that our panel of experts concluded that there was no convincing evidence for cold fusion. He gave me some polite excuses for ignoring our report and graciously accepted it. In his very next article he did mention our report and wrote a somewhat more balanced piece with the lead "Cold Fusion Gets Cold Shoulder from Many, a Year After Findings."

In arriving in Salt Lake City and picking up a copy of *Fusion Facts* (Vol 1, number 9, March, 1990), I gained new insight into Bishop's standing in Utah. *Fusion Facts* is published by the Fusion Information Center, an independent corporation. This monthly newsletter provides, what is advertised as, factual reports on cold fusion developments. While scanning the March issue, I noticed an article entitled *"Wall Street Journal* Resurrects Cold Fusion." This article discussed Bishop's piece in the *Wall Street Journal* on March 2, 1990 which carried the title "Cold Fusion Research Dispels some Doubts", where work in several laboratories was cited. None of the research cited, however, was new and it had all been examined by our panel before issuing our final report. At the end of this article in *Fusion Facts* there was an interesting note. It welcomed Bishop and his prestigious paper to the fusion club. The note predicted that Bishop's article would enhance interest in cold fusion by private industries. It was also hoped that such journalism

---

26 The 1990 judges for the AIP award were Mr. Howard Lewis (Chair, Editor, National Association of Science Writers newsletter), Mr. Robert W. Cooke (Newsday, Inc.), Ms. Shannon Brownlee (US News and World Report); Dr. Clifford B. Swartz (SUNY, Stony Brook), Mr. Sanford Pelz (the Browning School) and Dr. W. Peter Trower (Virginia Polytechnic Institute and State University).

would catalyze funding in the United States so that researchers here would catch up to the funding levels provided scientists in Japan and India by their governments and corporations. Finally the note took a slap at the U.S. Department of Energy for its funding policy on cold fusion.

As late as February 7, 1991 Bishop was still echoing a positive impression of Fleischmann and Pons' cold fusion claims in his article "Cold fusion verdict may be delivered soon." Bishop quotes Pons' lawyer C. Gary Triggs as saying that two new scientific papers in preparation are "very, very important" and contain "some very exciting stuff." Although Bishop is careful to state that most scientists dismiss the Fleischmann-Pons claim as spurious, he wants the reader to believe lawyer Triggs' message that a miraculous recipe for cold fusion was about to reach the public. This signal was reinforced with quotes from outside experts when Dr. Pons refused to give any information about their excess heat experiments. For example, Bishop included a quote from Dr. Loren G. Hepler who had reviewed Fleischmann and Pons' experiments recently. Hepler is quoted as saying "This reviewer had the feeling that Pons was willing to tell all, but the patent lawyer restrained him." As discussed several times before, when Fleischmann and Pons' claims were about to be closely scrutinized by experts, the two University of Utah scientists stopped the procedure by appealing to the patent process.

Fusion Facts is a most unique tabloid of science, published in the corporate world to profit from cold fusion. Only positive claims on cold fusion are given any credence. The editor made a strong case for cold fusion by exploiting the statistical data showing that the initial contributions of Fleischmann *et al*. and Jones *et al*. were the most frequently quoted papers in all of science in 1989. Unfortunately, citation frequency is not necessarily correlated with scientific merit. All visible indications are that the single purpose of *Fusion Facts* is to promote cold fusion as a valid science. When titles of negative papers are listed, they often are accompanied by a quip to belittle the contribution. For example, when no excess heat was obtained in a report by a well-known research group, a statement was included to say that it was obvious the authors needed help in operating a successful Fleischmann-Pons experiment. Accuracy and historical perspective were often lacking as evidenced by a report crediting Steven Jones as the discoverer of muon-catalyzed fusion. The Fusion Information Center sent a letter to key members of Congress on August 8, 1990 offering an alternative to dependence on foreign oil. The letter stated that the Department of Energy had been mislead by the DOE/ERAB panel, "most if not all of whom are supported by DOE hot fusion funds." This is untrue and illustrates the type of journalism promoted by the Center. The letter went on to cite work in Taiwan, Japan, Hawaii and Los Alamos as the conclusive proof of the reality of cold fusion, and urged the chairs of the Senate and House

Energy Committees to schedule public hearings. Congress, however, seems content to leave this technology to Japan [*Bull. Am. Phys. Soc.* **35** 2177 (1990)].

On the evening of March 30, Norman S. Bangerter, Governor of Utah, gave an inspirational talk, at a buffet reception, to the participants at the First Annual Conference on Cold Fusion emphasizing the future potential of cold fusion to the economy of Utah. Dessert was provided by Mrs. Fields Cookies, whose home office is located in Park City, Utah. These events were in harmony with the inspirational spirit of the Conference taking place in the mecca of cold fusion. The hyperbole placed on various fragments of claimed evidence supporting cold fusion excited the attendees, largely believers, at selected times to clap, scream or give a standing ovation. The fervor of proponents' belief in cold fusion can be best illustrated by a twelve bottles of wine to one bet made by Eugene F. Mallove at the First Annual Conference on Cold Fusion. Mallove bet Douglas Morrison that cold fusion will be demonstrated to be an *excess-energy generating nuclear phenomenon* (defined as 20 watts cm^{-3} of palladium) by December 31, 1992. If it is so demonstrated by said date, Morrison owes Mallove one bottle of wine. If it is *not* so demonstrated Mallove owes Morrison twelve bottles of wine. I am holding the written bet and have been designated to act as judge. Mallove exemplifies the passion of a true believer with his twelve bottle to one bottle bet at a time when there was "no convincing evidence for the phenomenon called cold fusion." All of the excitement and support for cold fusion at the Conference played well in Salt Lake City and I'm sure pleased the Fusion/Energy Advisory Council, most of whom were registered for the conference. I must admit that I was surprised that the conference brought out so many believers one year after the original press release.

Was the optimism and excitement displayed by the believers at the First Annual Conference on Cold Fusion justifiable based on the overall summation of experimental evidence produced during the one-year time interval following the March 23, 1989 press conference? The answer is a categorical no, for the following reasons.

(a) The major fraction (estimated to be 80% percent or more) of the world's experimental results, including the most comprehensive and best experiments, were negative. These negative results were excluded from the conference giving one a totally distorted picture of the current state of cold fusion.

(b) The experimental papers at the Conference presented no convincing evidence for cold fusion. In fact most of the claims presented were based on previous reports, some of which were several months old. Many of the more exotic early claims could either not be repeated again or were scaled down in magnitude such that the new signal approached or was of the order of the background. If the cold fusion phenomena were real, one would have expected a steady increase over the year in the number and quality of positive results.

(c) Serious questions about the accuracy of the cell calibration continued to plague those reporting excess heat. At the Conference, researchers from General Electric, who had a collaborative arrangement with the University of Utah on cold fusion research, challenged Pons' claim of excess heat (now at the 10 to 30% level compared to the early claim of several hundred percent) on the basis of their cell calibration. L.J. and T.F. Droege, initially strong proponents of cold fusion, confirmed that serious problems exist with cell calibration. In their paper they stated "There are enough calibration runs which show too much heat and D$_2$O runs which show little or no heat that the whole process could be noise."

(d) Most importantly, to this date there has not appeared a *single report* on cold fusion showing a consistent set of experimental data with commensurate intensities of excess heat and nuclear products. On this basis alone, one can say that at this time there is no evidence that the claimed excess heat, if real, is associated with a nuclear process.

The underlying assumption of the theoretical papers at the conference was that cold fusion was an *experimental fact* and took place under the unusual conditions that were available in metallic lattices such as those of palladium and titanium. The atomic-lattice environment was asserted to be of great importance and the key to cold fusion. The task undertaken by the theorists then was to explain how cold fusion actually happens in a way that was consistent with reported experimental claims, regardless of whether or not the claims were confirmed or even statistically justified. One of the theorists who gave a lecture at the conference was Julian Schwinger, a Nobel Laureate. His starting premise was that both the nuclear reactions D+D and p+D were responsible for cold fusion. His theory was formulated in a way to suppress the ^{3}He+n channel of the D+D reaction and at the same time to produce tritium and protons with very low kinetic energies. Hence, Schwinger's model created a source of tritium and suppressed the yields of any 2.45– and 14–MeV neutrons, consistent with the exotic claims of the BARC and Bockris groups. The reaction p+D was treated by Schwinger in a way to produce ^{3}He while highly diminishing the intensity of the conventionally observed 5.5–MeV gamma rays by coupling the reaction energy directly into the atomic lattice (see Chapter VIII). Schwinger commanded great respect and the organizers used his presence to promote and to authenticate cold fusion. The question period following Schwinger's talk displayed the fundamental weaknesses in his model. For example, skeptics attempted to have Schwinger confirm that tritium and ^{3}He were final states in his model, as appeared to be the case from his lecture. The sixty-four dollar question that followed was why are the yields of these products orders of magnitude less than the claimed amount of excess heat? This important question was left unanswered at the conference.

The most powerful voice defending cold fusion at the conference was that of Italian theorist Guiliano Preparata. He spoke often and usually with a degree of passion unmatched by the other proponents of cold fusion. Preparata proposed a model which he claimed not only 'explained' the smorgasbord of positive results that had been reported to date but also 'explained' why certain groups with sophisticated equipment consistently obtained negative results. His grandiose, but radical, scheme to alter nuclear physics in solid lattices is contrary to the known small effect of the chemical environment on nuclear decay processes (see Chapter VIII). Examination of Preparata's assumptions, even at the most elementary level, raises serious questions about their validity. His 'explanation' of the large stockpile of negative results, for example, was that these come from experiments where the D/Pd ratio had not reached the obligatory threshold of one. If such a threshold were necessary, essentially all the positive claims of cold fusion would be automatically eliminated, since most of these were carried out at values considerably below one. In summary, the 'theories' or models proposed to date to 'explain' cold fusion are very unsatisfactory. In fact, no logically coherent theory exists and many skeptics felt that such a theory will likely not be forthcoming (at the present time there is not even a reproducible set of experimental data for models to fit).

In the past, proponents of cold fusion had often targeted the 'Eastern Establishment' and the DOE/ERAB panel as the major villains whose actions were detrimental to the growth and development of cold fusion. At the conference other villains emerged, namely the journal *Nature* and its Editor, John Maddox, and Associate Editor, David Lindley. Both the journal and its editors were frequently ridiculed from the podium or floor of the meeting. Maddox and Lindley had each written scathing editorials that appeared in *Nature* during the week of the conference. An article critical of the press was also written by Fleischmann and Pons (published in the *Deseret News*) and distributed at the conference. Although Fleischmann and Pons stated their annoyance with the press in general (from this observer's vantage point it appeared that the press had given them fantastic coverage), they singled out *Nature* and its editors as targets to gore. Much of their displeasure stemmed from *Nature's* publication of several articles with negative results, while rejecting those articles that were received with positive results including their own. From my experience with several of these papers, I commend *Nature* for its handling of these manuscripts through the normal peer-review process.

Following the revelation that the University of Utah had paid attorney C. Gary Triggs (recall that he had written threatening letters to faculty) $68,000 in fees, Gary Taubes and Tim Fitzpatrick discovered in May, 1990 that a reported $500,000 gift to the National Cold Fusion Institute (NCFI) had actually come from a University of Utah fund. This large sum of money was advertised to prospective donors to cold fusion research at the NCFI as an anonymous donation, whereas in fact the money had come from the University of Utah Research Foundation through the office of University of Utah President, Chase N. Peterson. The disturbing sequence of events at the University of Utah, culminating in the misidentification of $500,000 as a gift from an unnamed source, prompted Hugo Rossi, dean of science, and twenty-two science faculty members to demand a complete scientific review and financial audit of the NCFI. The administration and the Fusion/Energy Advisory Council approved these reviews, on which I will comment later.

The NCFI had internal problems of its own. Rossi had remarked that one of these problems was a lack of openness among all the scientists at the institute. Furthermore, while he was interim director he was wisely attempting to have the institute demonstrate "in a convincing manner" whether the Fleischmann-Pons phenomenom was real and reproducible. However, this goal, Rossi said, "was contrary to the objectives of Pons and Fleischmann" who were "more interested in producing a demonstration device which convincingly and consistently gave off excess heat" (*The Chronicle of Higher Education*, November 14, 1990). Rossi eventually resigned as interim director of the NCFI. After Dr. Will became the first permanent director of the Institute, some of the physics programs were directed again towards establishing the existence of cold fusion as Rossi had originally attempted. Other programs, however, were less structured and plagued by problems of direction and interpretation.

Besides Peterson's misrepresentation of the $500,000 NCFI fund and his lack of prompt action in defending Salamon and other Utah faculty when they were threatened by attorney Triggs, Peterson had come under sharp criticism also for other transactions. He had been forced to return a donation of $15 million in pharmaceutical stock when faculty and students learned that this gift to the university was contingent on renaming the medical school and hospital after the donor, James L. Sorenson. Many on campus, including the physics faculty, continued to be upset by the circumstances surrounding the March 23, 1989 press conference. The overly optimistic tone of the press conference misled the public to believe that very simple experiments had demonstrated the feasibility of room temperature nuclear fusion which would produce an inexhaustible supply of clean and cheap energy. Even the facts stated at the initial press conference were wrong. It was announced that the experiment had produced four times as much energy as it consumed, a statement completely inconsistent with the data in Fleischmann and Pons' first paper which was submitted for publication several days before the press conference.

Peterson and his administrative assistants had made the decision to go public with the Fleischmann-Pons claim without either an adequate internal or external review. It was highly inadvisable to go public with a cold fusion claim without seeking advice from nuclear physics experts in the local physics department. Peterson himself had described the Fleischmann-Pons claims to Professor Hans Bethe, distinguished nuclear theorist at Cornell University. Bethe had urged caution and recommended that the announcement be delayed. However, Peterson chose to ignore Bethe's advice. Very soon after the press conference, Peterson had arranged a Congressional hearing (see Chapter V) promoting the claim before any confirming evidence was available. Dr. Robert Park, Executive Director of the American Physical Society, had remarked that the University of Utah's financial involvement in the cold fusion enterprise was inexcusable because "Pure greed was apparently the only motivation." Over the months following the press conference, the controversy on campus had grown and the faculty had become more frustrated because of the administration's unflagging support for the Fleischmann-Pons cold fusion.

On June 1, 1990 Peterson admitted that hiding the origin of the $500,000 had been a mistake. Faculty criticism continued and on the afternoon of June 4 Sandra Taylor, a professor of history, introduced a resolution to the Academic Senate, which read in part:

> The academic senate respectfully requests that the Institutional Council and the Board of Regents examine the question of whether continuation in office of the current president is in the best interest of the University of Utah and the community which it serves.

The resolution was seconded and passed. The discussion indicated the faculty's concern about the president's continued support of cold fusion and "argued that his actions had damaged the university's reputation because of the overwhelming scientific evidence indicating that the purported phenomenon does not exist" (*The Chronicle of Higher Education*, November 14, 1990). On June 11 Peterson, who was sixty years old and president since August 1983, announced that he would take early retirement at the end of the 1990-91 academic year. He said his action was intended to (*ibid.*):

> [R]educe the present level of unproductive controversy on campus so we all can focus on the enormous strengths and contributions of faculty, staff and students.

With the pending reviews of the NCFI and Peterson's announced retirement, campus tempers cooled for the time being.

The number of surprises surrounding the cold fusion saga continued to increase when two days before a scheduled meeting of Utah's Fusion/Energy Advisory Council on October 25, 1990, local media noticed that Pons was missing. The council had been expecting either Pons or Fleischmann to present their recent data in preparation of the council's deliberations on the state of cold fusion before releasing the final approximately one-million-dollar installment of the state's cold fusion fund. Fleischmann was at home in Southampton, England, receiving medical treatment and claimed that he had not been notified of the meeting. Pons' whereabouts seemed a mystery. His home was up for sale and his telephone was disconnected. Even University of Utah Administrators didn't know, or weren't releasing, his whereabouts. Friends and neighbors thought he was in France. Fritz Will, the Director of the NCFI, insisted that all efforts had been made to notify both men of the important meeting. It had become abundantly clear that the communication between Will and the two cold fusion gurus was very limited, and essentially nonexistent with Pons. Communication with Pons was possible only through his attorney and Will refused to discuss scientific questions through a lawyer. On October 24, 1990 Pons was heard from when the university received a one-sentence fax message from Pons' attorney Triggs requesting a sabbatical leave (which was not granted). Later Utah Assistant Attorney General, Joseph Tesch, spoke with Pons through a conference call arranged by Triggs. Pons promised to return to the university on November 7 to participate in the scientific review of the NCFI, a study demanded several months earlier by a group of faculty [*Science* **250** 755 (1990)].

The scientific review of the NCFI finally took place on November 7, 1990 with Pons in attendance. The review committee consisted of Professors Robert K. Adair, Stanley Bruckenstein, Loren Hepler and Dale Stein. The charge to this committee has not been published. According to media reports, however, the task of the committee was to examine the quality of the research being performed at the state-funded NCFI. Fusion/Energy Advisory Council member Hansen, one of the two scientists on the nine-member council, said that "this is not a review to see if cold fusion is real" (*C&E News*, November 5, 1990). He went on to explain that the committee would try to decide if public money had been spent on properly conducted, defensible research. If this was all the four-person committee was to give advice on, the Utah Advisory Council would have the option to release more state funds even though none of the NCFI programs had presented any convincing evidence for the existence of cold fusion.

The four external review committee members were asked to sign an agreement restricting their freedom to discuss any review materials with others. This was in accord with Will's statement made months earlier that he would recommend that these reviewers be required to sign statements agreeing not to disclose information of a proprietary nature that would endanger the Utah patent applications [*The Chronicle* **30** A15 (1990)]. This restriction of scientific information due to security imposed by patents had raised its ugly head several times previously when Fleischmann and Pons' claimed results came under close scrutiny. So in this sense, withholding material in the present review was nothing new. What was new, however, was that by now many faculty members were questioning whether the university should continue to support cold fusion research and allow state funds to finance it.

The concept of a National Cold Fusion Institute was born within the first weeks of the March 23, 1989 press conference, when Utah State and University administrators actively sought state and federal funds for cold fusion research (see Chapter V). It is not surprising, therefore, that most of the institute personnel believe that the Fleischmann-Pons phenomenon of excess heat has its origin in a room-temperature nuclear fusion process. The architects of cold fusion at Utah, Fleischmann, Pons, Walling and Simons, however, adopted a very exotic non-conventional view of nuclear physics, exemplified by the Walling-Simons 'model' which I described in Chapter III as a three-miracle 'non-theory'. These miracles, claimed to be applicable to cold fusion in atomic lattices, were created to alter both the fusion rates and branching ratios of the D+D reaction, as well as to hide high-energy gamma rays for fusion capture channels. Many of the programs at the NCFI reflect this Utah philosophy where no specific nuclear fusion model is tested and the fusion signals are indefinite and ambiguous. This situation makes these programs vulnerable to mistakes, especially when non-correlated signals are misinterpreted. The program of the physics group directed by Bergeson is an exception. Although this group accepts the rate enhancement miracle, at the level of the Jones BYU group, they otherwise base their experiments on conventional nuclear physics signals. Bergeson was previously a member of the Salamon group that examined the Fleischmann-Pons electrolytic cells and found no evidence of fusion by-products.

Advocates of cold fusion have assembled long lists of research groups claiming positive results. Scientists at the NCFI have assembled such a list of ninety groups, using it as evidence to justify their claim of a new nuclear phenomenon. Bockris and the editor of *Fusion Facts* have compiled similar lists (although there are differences in the three lists, for example, these lists include claims from ten, twelve and sixteen countries, respectively). The sources of these claims come from publications, internal reports, letters or private communications. The number of claims in the lists published in main-line, peer-reviewed journals are very few. Anyone familiar with the history and quality of the claims can strike case after case from these lists, either because the authors have retracted their original claim or the claim is unsubstantiated and not reproducible. The lists include also some claimed positive results from the NCFI, where small amounts of excess heat and tritium have been reported. These claimed positive results add to the length of the list, but none of these results are definitive and reproducible. It would be to everyone's benefit if the proponents of cold fusion would place less emphasis on counting numbers of positive claims, and more emphasis on establishing the validity of even one claim, where two signals (e.g. excess heat and fusion products) are understandably related.

The committee of four scientists from outside Utah, appointed to evaluate the credibility of the scientific work at the NCFI, concluded "that neither the institute's various experiments nor those anywhere else have firmly established the existence of cold fusion." This conclusion is the same as that reached by our DOE/ERAB panel some one and a half years earlier. The fact that over this long time period no progress has been made on solving the highest-priority task of establishing whether or not the phenomenon called cold fusion is real, raises serious questions about future research directions and funding patterns of cold fusion. At the time of our DOE/ERAB panel's final report, we estimated that several tens of millions of dollars had been spent in the United States on cold fusion. The world's scientific institutions have probably by now squandered between $50 and $100 million on an idea that was absurd to begin with. The question then to be addressed by Utah's Fusion/Energy Advisory Council was how much, if any, of the remaining approximately one millon dollars of state money should be released for continued research at the NCFI? The Advisory Council answered this question by releasing the final installment of the state's $4.5 million appropriation to the NCFI. By cutting its staff from twenty-five to seventeen, these remaining funds allowed the Institute to operate until June 30, 1991. The emphasis of the institute's program during its last gasp for survival was still on the controversial claim of Fleischmann and Pons that deuterium nuclei fuse in a palladium electrode, thereby releasing heat. There is still no convincing evidence for such a process!

On January 9, 1991 the University of Utah announced the resignation of Pons from his tenured faculty position, effective on January 1. He was appointed as a research professor for a limited period of eighteen months, until June 30, 1992. The Utah Advisory Council gave Pons a deadline of February 1 to turn over all of his research notebooks to Council member Hansen. The affiliations of Fleischmann and Pons with the NCFI were severed and future state funding of their work was tied to their cooperation in having others verify some of their claims.

Patents have played a major role in the cold fusion saga. Secrecy has restricted outside collaboration and has kept Fleischmann and Pons' research from close scrutiny. For example, the outside review committee that met on November 7, 1990 to evaluate the NCFI program had to sign a confidentiality agreement. Even so, Adair stated that "important details of the Pons-Fleischmann work were withheld from the committee" and "until their programs are openly described, and independent confirmation is possible, those programs cannot be judged scientifically." The motivation for Fleischmann and Pons to withhold information is particularly puzzling when patent office libraries around the world are publishing full details of their patents (*New Scientist*, November 10, 1990). Any member of the public can examine this material in any patent library or purchase it outright as I have done.

Utah's international patent application was filed March 12, 1990 and is based on seven applications filed in the United States between March 13 and May 16, 1989. One of the surprising things about the list of inventors is that in addition to Fleischmann and Pons, Walling and Simons are included as senior participants in this University of Utah episode. They were included to meet the requirements of the U.S. Patent Office that everyone who has contributed to an invention is credited. Walling and Simons wrote and published a theory "explaining" cold fusion which I characterized as a three-miracle 'non-theory'. I, for one, find it hard to imagine how such a totally erroneous model can contribute in any meaningful way to a patent application. In fact, one would expect such supplementary material to be a strong deterrent.

The international patent application is filed under the World Intellectual Property Organization's Patent Cooperation Treaty and designated for all forty-two member states, including, for example, Japan, Monaco, Chad and the United States, however, excluding India and Mexico. The Abstract states:

> The present invention involves an apparatus and method for generating energy, neutrons, tritium, or heat as a specific form of energy. The apparatus comprises a material such as a metal having a lattice structure capable of accumulating isotopic hydrogen atoms and means for accumulating isotopic hydrogen atoms in the metal to a chemical potential sufficient to induce the generation of the specific items. The sufficient chemical potential is, for example, enough to induce the generation of an amount of heat greater than a joule-heat equivalent used in accumulating the isotopic hydrogen atoms in the lattice structure to the desired chemical potential.

Those familiar with the March 23, 1989 press conference and all the following publicity generated by the University of Utah will notice that the now household words "cold fusion" do not even appear in the abstract. The focus of the abstract is on the special role performed by the chemical potential in the generation of heat, neutrons and tritium from metals loaded with isotopic hydrogen atoms. This emphasis on the chemical potential is completely consistent with Fleischmann's many statements about the enormous pressures he envisioned would be attained in the palladium cathode during electrolysis. It is this central, but flawed, idea of Fleischmann that led him and Pons to embark on their early experiments. Elsewhere in this manuscript I've described the rather modest pressures attainable by electrolysis and the fact that these pressures play an insignificant role in fusion.

As I guess one would have expected, the Utah patent application is extremely vague about the crucial parts of the process. The seven United States patents are in three general areas (the filing dates all in 1989 are included).

Area 1 – Heat Generating Method and Apparatus; March 13, April 10 and April 14.

Area 2 – Neutron-Beam Method and Apparatus; March 21.

Area 3 – Power Generating Method and Apparatus; April 18, May 2 and May 16.

The international application follows very closely the above areas and is divided into three sections.

Section 1 – Heat-Generating Conditions Within a Metal Lattice.

Section 2 – Neutron Generation and Applications.

Section 3 – Detailed Analysis of Enthalpic Heat Production.

The international document of over 100 pages, including figures, does not contain any new secret information about how cold fusion occurs. The vagueness of the document may be illustrated by the following quotes:

One possible mechanism for the nuclear-fusion events which are believed to occur within a metal lattice charged with isotopic hydrogen involves a correlation between the valence electrons in the metal lattice and pairs of isotopic hydrogen which allows the hydrogen-atom pairs to become more localized and therefore more likely to fuse. Fusion occurring according to this mechanism may be favored by fermionic metals . . .

The most surprising feature of these results is that the enthalpy release apparently is not due to either of the well established fusion reactions . . . the enthalpy release was primarily one which is aneutronic and atritonic.

It should be noted that the present data support the view that the 'steady-state' enthalpy generation is due to a process or processes in the bulk of the electrodes although this statement does not now seem to be as clear cut as that made on the basis of the results contained in the inventors' preliminary publication.

In a very spectacular statement about future possibilities of energy gain the inventors claim "reasonable projections to 1000% can be made"! This is to be compared to 300%percent (four watts output for one watt input) claimed at the original press conference and 10 to 30% claimed in the original and subsequent publications (these latter data are those included in the patent application).

Section two of the international patent document deals with neutron generation and applications, including a neutron-beam generator, neutron radiography, neutron diffraction and radiative neutron capture. All of these applications require an intense source of neutrons! It is highly ironic to include such a neutron application section in the patent application along with Section three where it is stated that the excess heat is apparently not due to the well established reaction $D+D \rightarrow {}^3He+n$ because the source of the heat was primarily aneutronic (as well as atritonic). The only evidence for invoking a nuclear process was the claim that the magnitude of the excess energy was so large "that it is not possible to ascribe this to any chemical process." In light of the recent admission of the NCFI that the scientific reality of cold fusion has still not been established at either the NCFI or elsewhere, one wonders how such negative publicity will impact on the Patent Examiners.

On June 30, 1991 the National Cold Fusion Institute (NCFI) closed its doors after approximately two years of operation. The five million dollars from the state of Utah had been spent and no additional monies were forthcoming from federal agencies and private sources. The NCFI's ability to raise money was damaged by the scandal associated with the transfer of 500,000 dollars in university funds to the Institute by former University of Utah President Chase Peterson, disguised as an anonymous donation.

On July 15, 1991 the University of Utah issued a news release announcing that the NCFI had issued a final three-volume report to the State Fusion/Energy Advisory Council. The news release painted a very positive picture of the current state of cold fusion stating "[O]ne hundred and thirty groups in fourteen countries have now reported various types of evidence of nuclear reactions in deuterium-loaded metals or compounds . . . [T]his includes evidence for excess heat, tritium, neutrons, x-rays, or gamma rays, and helium or charged particles." Such optimism may have been necessary to placate the Advisory Council after two years of financial support. The bottom line however, was that the institute scientists were not able to confirm Fleischmann and Pons' claim that watts of heat due to nuclear fusion at room temperature occurred during electrolysis of D_2O in a simple cell with a palladium cathode. Not a single experiment has reported consistent results where the yield of fusion products was commensurate with the claimed excess heat. Utah's dream of a fusion valley had evaporated as evidenced by the fact that the state was unwilling to put up more money. In summary, five million dollars of state funds were squandered on a research program that was based on a far-out idea which had not been adequately peer reviewed.

XI

Cold Fusion and Polywater

The cold fusion affair exhibits similiarities to the polywater episode of the 1960s and 1970s. To have had two major episodes in science, such as cold fusion and polywater, occur in the same half century is noteworthy. Both of these episodes attracted wide attention in the popular press. Both cold fusion and polywater had the attractive feature that experiments appeared to be very simple and required only elementary equipment that could be found in an advanced high school or college freshman laboratory. In contrast to space research and particle physics which require a battalion approach, a one- person or a few-person team could enter the new fields of polywater and cold fusion research and hopefully make significant contributions in a short time. Hence, quick and premature publication was one of the characteristics of work in both areas. Due to advanced communication methods, cold fusion results appeared more rapidly and more often in the press.

Anomalous water, later assigned the name of polywater, was first reported in 1962 by N.N. Fedyakin who worked at the Technological Institute in Kostroma, USSR. Soon after Fedyakin's results were published, all future work in the USSR on anomalous water was under the direction of B.V. Deryagin, working at the Institute of Physical Chemistry in Moscow. Little is known about Fedyakin's later work except that his name appeared in print only in conjunction with members of Deryagin's group. Deryagin, born in 1902 and an eminent scientist and corresponding member of the Soviet Academy of Sciences, directed a large group studying the nature of forces between solids. It is he who was the spokesman and leader of all polywater research in the USSR.

Although a few western scientists knew about anomalous water, even fewer understood its significance prior to the meeting of the Faraday Society in 1966. Deryagin's paper at this meeting contained a bibliography of thirty-six previously published manuscripts on polywater, thirty of these originating in the USSR and of the latter, twenty-three have Deryagin as a co-author. One of the startling conclusions of Deryagin's paper at the Faraday Society Meeting was that anomalous or polywater was claimed to be the stable form of water and that in time all water would turn into the anomalous form, even though the conversion may be a slow process. Anomalous water was postulated to be formed in two steps. First, by quick condensation when the vapor was supersaturated, and secondly, on decreasing the vapor pressure, anomalous water was formed slowly by condensation in a subsaturated atmosphere. Although polywater was produced from ordinary water, it did not freeze or boil like ordinary water and it hardly resembled water in any of its properties. Some described polywater as a viscous substance like vaseline.

The early history of polywater and cold fusion developed in quite different ways. The polywater furor reached epidemic proportions in the United States only after a widely publicized article by E.R. Lippincott et al. appeared in Science [**164** 1482 (1969)]. This work, a joint effort of scientists at the University of Maryland and the National Bureau of Standards, reported the results of infrared and Raman spectroscopic measurements of anomalous water. On the basis of their interpretation of the spectral data, Lippincott et al. proposed that anomalous water was a true high polymer, consisting of ordinary water monomer units. "The properties, therefore, are no longer anomalous but rather, those of a newly found substance – polymeric water or polywater." This is the origin of the name polywater which largely replaced the early designation of the material as "anomalous" water. The above authors proposed structures which explained the remarkable properties and stability of the new material which they claimed to be a true polymer of water and justified the name polywater.

Lippincott and co-workers received wide publicity for their structural work on polywater and this publication sparked furious activity and heated debate among scientists. All of this occurred several years after the discovery of anomalous water in 1962 by the Russian scientist N. Fedyakin. Most of the early work by Deryagin and his many co-workers went essentially unnoticed by scientists in the west. Following the Science article, publications on polywater increased rapidly, reaching a peak in 1970, some eight years after the discovery. Once scientists were aware of Lippincott's paper on polywater, in which he reported spectra that were very different from ordinary water, research on polywater caught the imagination of many in the United States and in other western countries. Lippincott's name became strongly linked with that of Deryagin in the polywater issue.

In contrast to the long delay prior to the zenith of the polywater affair, cold fusion research peaked within the first year after the initial press conference. The rise in interest among scientists and the media in the two episodes are more parallel if one delays the start of the polywater episode to Lippincott's highly publicized article that appeared in 1969, seven years after the discovery. Some 500 papers were published on the subject of polywater over the period 1962 to 1974, of these over 90% were published during or after 1969. Approximately 15% of the total publications originated from the USSR, 45% from the United States and 40% from elsewhere. During the first ten months after the announcement of cold fusion, 141 scientific papers have been published on this subject (*New Scientist*, April 21, 1990). Most of these articles discuss the original paper of Pons and Fleischmann published in April, 1989 in the *Journal of Electroanalytical Chemistry*. The experimental papers on cold fusion over the first ten-month interval were more than four to one against the phenomenon. As already stated, the history of polywater and cold fusion are somewhat similar if one traces polywater from Lippincott's highly publicized 1969 paper. The long delay from 1962 to 1969 is understandable since anomalous water during this period was almost unknown and little understood in the west. Once polywater was made respectable in 1969 by articles in leading newspapers and news magazines (e.g. the *New York Times* and *Newsweek*), polywater developed in much the same way as cold fusion did two decades later. The somewhat more rapid escalation of cold fusion activity is probably the result of a number of factors, including the current widespread use of BITNET and FAX machines. The widely publicized media message that cold fusion would solve the energy crisis for all time contributed also to the early frenzy. Another factor was that early on Admiral Watkins, the Secretary of Energy, mobilized the efforts of many scientists in each of the ten large national laboratories to study the cold fusion phenomena. In addition, the study of cold fusion also caught the imagination of university and industrial scientists immediately after the March 23, 1989 press conference, both due to its simplicity and its unlikely occurrence.

A case can be made that the phenomena known as polywater and cold fusion are atypical in that neither polywater nor cold fusion were the preserve of any one scientific specialty. Polywater, initially the domain of surface chemists, attracted chemists from other areas, physicists and life scientists with many different backgrounds and research specialties. The same was true for cold fusion, where the initial experiments involved electrochemistry. Very soon, however, a number of scientists from disciplines outside chemistry became involved, including physicists and materials scientists. Early on, it was evident to some, but not all, that a collaboration of specialists from several disciplines was essential for a successful experiment in cold fusion. The close collaborations were even more important in cold fusion than they were in the study of polywater. The fact that several different disciplines participated made the debate more stimulating and competitive but also more controversial.

Polywater and cold fusion were atypical also in another sense. One of the characteristics of real discoveries is that they are often made by a number of independent groups fairly near in time. Similar lines of forefront research are usually being conducted by more than one group, such that confirmation of a discovery is soon established by the other related groups. Neither in the case of Fedyakin's claim of polywater nor in the case of Fleischmann and Pons' claim of cold fusion were other groups able to satisfactorily reproduce their reported results. In the case of the discovery of high temperature superconductivity, for example, many groups were able to confirm the claim in a relatively short time. Cold fusion and polywater were plagued by irreproducibility, inconsistency, backgrounds and impurities.

It is particularly intriguing that the two major scientific episodes of this century both involved water, one of the commonest materials known to man. In the case of polywater, the claim was made that this strange new substance with unusual properties could be formed from ordinary water. Speculation about its importance and possible applications were wide spread, even though solid evidence for its existence was lacking. It was considered in the media as something of potential economic importance. The *Wall Street Journal* articles were particularly sensational. All of the same could be said of cold fusion. In this case deuterium, a heavy isotope of hydrogen present in water, was claimed to provide an unending supply of energy by fusing at room temperature in a simple electrolytic cell. The *Wall Street Journal* broke the story on March 23 and 24, 1989 in the United States, with articles entitled "Development in Atom Fusion to be Unveiled" and "Taming H-Bombs."

Cold fusion and polywater were also very similar in terms of media coverage, each was probably more extensively covered than any other science event during the twentieth century. It is understandable in terms of the striking, but controversial, claims that were made and the possible implication of the claimed discoveries for society as a whole. Typical headlines from the two eras were "Weird Water: Debate Over Mysterious Fluid Splits Scientific World: Polywater Has a Bright Future – or Does It?" and "Fusion Breakthrough Ignites Scientific Uproar." Each of these titles illustrates the active scientific debate and controversy going on in the scientific world as projected on a daily basis by the media. During the polywater years, Deryagin and Lippincott were household names. The same can be said for Fleischmann and Pons during the cold fusion era. Although the media stories were not always particular about the technical details, the various writers tended to be familiar with the economic, political and sociological aspects of the scientific issues. Most of these science writers were skilled at describing sometime complicated scientific concepts in terms that the interested lay reader could appreciate and understand. Many of the stories were embellished with interviews or quotes from scientists, sometimes, but not always giving opposing views.

Much of the polywater research went on with no true collaboration between research groups of different persuasions. Franks [(*Polywater* MIT Press (1981)] illustrated the great disadvantage of these individual groups working in isolation when he stated:

Had such an exchange taken place, then independent investigations on common samples would soon have shown up the possible effects of impurities on the recorded spectra. As it was, any such doubts were always countered by the claim that the doubter had not yet perfected the proper techniques for the preparation of pure polywater. Thus, the uncertainties could have been resolved much sooner . . .

The pattern was exactly the same in cold fusion. Those who were not able to produce excess heat were dismissed as not having the proper type or dimension of Pd cathode, improper current density, too short an electrolysis time, etc. Those claiming heat either refused or were very reluctant to collaborate with experts in particle detection. Again many of the discrepancies could have been resolved much quicker by close collaboration between believers and skeptics.

Other similarities exist between the cold fusion and polywater episodes when one compares the statements of advocates of the two different phenomena. John D. Bernal, well respected British scientist, stated in an interview that in his opinion polywater was the most important physical-chemical discovery of this century. This is to be compared to a statement about cold fusion by Cheves Walling, a colleague of Stanley Pons at the University of Utah and co-inventor of the phenomenon. Walling stated that "the concept (of cold fusion) could be the most exciting development of his lifetime, exceeding the Manhattan Project." Speculations on the applications of polywater and cold fusion were equally imaginative. Both phenomena also evoked warnings about their danger. F.J. Donahoe attracted much media attention by his small article in *Nature* [**224** 198 (1969)] about the possible dangers of polywater. His article concluded by stating that:

Every effort must be made to establish the absolute safety of the material before it is commercially produced . . . scientists everywhere must be alerted to the need for extreme caution in the disposal of polywater. Treat it as the most deadly virus until its safety is established.

In their initial paper Fleischmann and Pons also issued a strong warning:

We have to report here that under the conditions of the last experiment (Pd-cube electrode), even using D_2O alone, a substantial portion of the cathode fused (melting point 1554°C), part of it vaporized, and the cell and contents and a part of the fume cupboard housing the experiment was destroyed.

Fleischmann and Pons concluded that a plausible interpretation of their results was in terms of ignition (i.e. D+D fusion). Although the University of Utah made much publicity about the incident of the vaporization of the palladium electrode, no one was in the laboratory to observe what had happened. The only evidence was the after effects. Hence, it was not possible to rule out the possibility that the incident was due to a chemical explosion rather than due to nuclear fusion.

A surprising number of theoreticians became deeply intrigued by and committed to the episodes of polywater and cold fusion. In the case of polywater, for example, the existence of a new form of water was in no way established before people started constructing theoretical frameworks and models describing how the stability of the unusual material was to be interpreted. Right from the beginning the real and most important issue was not whether the claimed phenomenon was real and correct, but how the reported data and claims could be explained. However, the commitment to the theoretical explanation was expressed and carried out with a wide range of intensity. An intermediate position on the theory of polywater was expressed ably by Keiji Morokuma, a colleague of mine at the University of Rochester during the polywater episode. He stated [*Chemical Physics Letters* 4 358 (1969)]:

> [T]hough the existence of 'the anomalous water' is still questioned and will take time to be established and models proposed are not concrete, it would be fruitful to begin to examine from a theoretical point of view whether such models could be of any possibility.

An example of a theoretician who committed himself completely to the existence of polywater is Leland C. Allen of Princeton University. He and Peter A. Kollman published a paper entitled "A theory of Anomalous Water" in *Science* [**167** 1443 (1970)]. These authors stated:

> [I]n this article we attempt to provide an independent and reasonably quantitative theoretical reference point for assaying the likelihood that the new form of water originally reported by Deryagin *et al.* exists, with the properties suggested by Deryagin and by the recent spectroscopic measurements of Lippincott *et al.* The stability, bonding, geometry, spectroscopy, formation, viscosity, and high- and low-temperature behavior of this substance are discussed on the basis of many-electron quantum mechanical calculations.

The paper by Allen and Kollman was a monumental effort to explain the claimed properties of a substance that did not exist. The assumptions and approximations inherent in their very complex quantum-mechanical calculations coupled with the primitive techniques available at that time allowed Allen and Kollman a wide range of possibilities extending from either complete support of the existence of polywater to its nonexistence. These authors chose to take a positive view since they could construct what appeared to them to be a self-consistent picture of a most unusual material.

During and prior to the Allen and Kollman paper, many reports were circulating that the eccentric properties of the substance initially known as anomalous water resulted from impurities in ordinary water. It is amazing that Allen and Kollman made no references to these disturbing signals. A common feature of most theoretical papers that can be characterized as attempts to explain how phenomena such as cold fusion and polywater might occur is that they seem unaware of analytical measurements showing impurities, backgrounds or other inconsistencies with their results. Allen and Kollman are to be credited, however, for changing their position in a short article in *Nature* [**233** 550 (1971)], even though it was for the wrong reasons. In this communication entitled "Theoretical evidence against the existence of polywater", the authors proceeded to use again complex computational calculations to take a position directly opposite that of their original work. Again no mention was made of the mounting evidence during this interval that polywater was nothing more than ordinary water with impurities, a factor that no doubt was the determining one for writing their last paper on polywater.

In the case of polywater the charge was made that theoreticians can explain anything if they want to badly enough, including why a non-existing substance was stable. Before giving some similar examples in the area of cold fusion, a word of caution is in order. It is important to keep in mind that many theorists chose not to become involved in either the polywater or cold fusion episodes for various reasons. Some because experimental verification was lacking, some because the stated claims were too preposterous to be taken seriously, some because a definitive prediction from established theory was either negative or impossible and they felt that the issue should be settled by experimentalists, and some for other more intuitive and personal reasons. An example of one who chose not to become involved in polywater is Joel Hildebrand, one of the great chemists of the twentieth century. He wrote in a letter to *Science* [**168** 1397 (1970)] that:

[P]roponents of polywater . . . may be interested to learn why some of us find their product hard to swallow. One reason is that we are skeptical about the contents of a container whose label bears a novel name but no clear description of the contents. Another is that we are suspicious of the nature of an allegedly pure liquid that can be prepared only by certain persons in such a strange way. We choke on the explanation that glass can catalyze water into a more stable phase.

A more exotic and analytical reason for rejecting polywater was given by Nobel-laureate Richard Feynman: "There is no such thing as polywater because if there were, there would also be an animal which didn't need to eat food. It would just drink water and excrete polywater" (and could use the energy difference to maintain its metabolism) [F. Franks, *Polywater*, MIT Press (1981), p. 157].

Theoretical 'explanations' of the cold fusion phenomenon appeared in record time. In this regard, the cold fusion phenomenon reached epidemic proportions on a much faster time scale than did polywater. Within three weeks of the press release announcing cold fusion, B. Whaley of the University of California, Berkeley, speaking to seven thousand chemists in Dallas proposed a boson screening theory of cold fusion. As discussed in Chapter III and elsewhere, theories explaining cold fusion propagated at a rapid rate with little or no concern for established concepts. Theorists explaining polywater based their efforts on the meagre evidence provided by a single spectrum while ignoring all the reports about impurities. In the case of cold fusion several theories were created to explain what was thought to be the production of ^{4}He by D+D fusion at room temperature in an electrolytic cell without regard for having such a bizarre result experimentally confirmed. The rush to obtain patents and establish scientific priority severely affected sound judgement.

In addition to postulating a theory based on the production of ^{4}He for which no experimental evidence existed, Peter Hagelstein's early theory also required that the 23.8–MeV gamma ray, known to accompany ^{4}He production, be highly suppressed and the reaction energy absorbed in the atomic lattice. The latter was required since the high energy gamma ray was not experimentally detected. The likelihood of this occurring was extremely small as discussed in Chapter III. The rush to establish scientific priority was epitomized by the theoretical paper of the University of Utah scientists, Cheves Walling and Jack Simons. Their paper was submitted on April 14, 1989, three weeks after the Fleischmann and Pons' press release, and published on June 15. The motive behind Walling and Simons' theory was to explain the early ^{4}He data of Pons which was shown almost immediately to be incorrect. Walling and Simons' most unusual theory required three 'miracles' to 'explain' the reported results of Fleischmann and Pons (see Chapter III). This is a striking example of theorists being too much in a hurry to become involved on the basis of very meager and unsubstantiated evidence supplied by a single faulty experimental measurement of helium which was shown within a few days to be completely erroneous.

As stated in a recent note (*New Scientist*, April 21, 1990) a majority of the theoretical papers on cold fusion up to this date were positive, constructing various improbable scenarios to explain how the phenomena might occur. A striking example is the theory of Julian Schwinger, Nobel laureate in physics, published in the following journals: [Z. *Naturforsch* **45A** 756 (1990); Z. *Physik Atoms Molecules and Clusters* **D15** 221 (1990)]. He begins his theory with the following ingredients.

(1) The claim of Pons and Fleischmann to have realized cold fusion is valid.

(2) But, this cold fusion process is not powered by the D+D reaction. Rather it is an H+D reaction, which feeds on the small contamination of D_2O by H_2O.

(3) The H+D $\rightarrow$ ^{3}He reaction does not have an accompanying gamma ray; the excess energy is taken up by the metallic lattice of palladium alloyed with D.

(4) The coupling with the palladium-D lattice that rapidly siphons off nuclear energy, as it becomes available, had previously acted to suppress the Coulomb repulsion between H and D, and, indeed, to overcome it with an energy of attraction that significantly ameliorates the effect of Coulomb barrier penetration.

Starting with this set of assumptions, it is obvious that the dominant fusion reaction produces neither neutrons nor high energy gamma rays (postulate 3). Both of these features are needed because the experimental evidence showed that neither neutrons nor 5.5–MeV gamma rays were observed in cold fusion experiments. Now the H+D reaction has been well studied and is known to produce ^{3}He and a 5.5–MeV gamma ray, but no neutrons. Postulates (3) and (4) which serve to hide the 5.5–MeV gamma rays are contrary to everything that is known about nuclear physics. Gamma rays of comparable energy are emitted from low-energy nuclear reactions in solid samples, e.g. the well-known thermal-neutron capture gamma rays. These have been studied quantitatively and are of importance in the operation of nuclear reactors. Hence, the probability that any processes in which capture gamma rays are totally suppressed in favor of direct conversion into lattice heat is extremely small. In addition, no ^{3}He has been found in cold fusion experiments!

Why bother with theoretical speculations, if the postulates are so improbable or the phenomena in question are very unlikely to even exist? Schwinger, however, was not deterred. In the second paper he summarizes "Evidence is presented for the assertion that an H-ion in a deuterided lattice encounters a relatively narrow Coulomb barrier before fusing to form ^{3}He." Starting with postulate (1) in the above paragraph, Schwinger proceeded to build a superstructure to 'explain' the Fleischmann and Pons reported results. Just as in the case of polywater, where theorists built castles on a single shaky spectrum (ignoring all the experimental evidence for impurities), Schwinger assumed the validity of Fleischmann and Pons' claim even though the heat and particle measurements differed by more than eight orders of magnitude and no helium and high-energy gamma rays were observed. Schwinger had submitted a paper on cold fusion to *Physical Review Letters*, one of the more prestigious physics journals. The paper was rejected with reviews by peers that were so harsh that Schwinger resigned from the American Physical Society.

Testimony to the nervousness of editors accepting theoretical papers on the subject of cold fusion is the editorial note that accompanied some of these papers. In the case of the Walling and Simon paper, the editorial comment is rather long and I include only the first two sentences:

> The preceding Letter contains new theoretical ideas which the Editors felt should be placed in the public domain, whether or not the experiments cited as evidence of 'cold fusion' are valid. For highly debated research, however, the Editors also believe that a summary of favorable and unfavorable comments by the reviewers should accompany the Letter.

This comment is most unusual and very revealing. The message is to publish the 'theory' even though it was constructed to 'explain' a reaction that may not exist. In fact it was known well before the publication of the Walling and Simon paper that the ^{4}He data, which the 'theory' (or better the set of assumptions) was formulated to 'explain', was totally incorrect. Still the paper was published.

The editorial note prominently displayed in box form on the first page of the second Schwinger paper states:

> Reports on cold fusion have stirred up a lot of activity and emotions in the whole scientific community as well as in political and financial circles. Enthusiasm about its potential usefulness was felt but also severe criticism has been raised. If in such a situation one of the pioneers of modern physics starts to attack the problem in a profound theoretical way we feel that it is our duty to give him the opportunity to explain his ideas and to present his case to a broad and critical audience. We do, however, emphasize that we can take no responsibility for the correctness of either the basic assumptions and the validity of the conclusions nor of the details of the calculations. We leave the final judgement to our readers.

Sociologists of science should in the future have a field day with these editorial comments when researching cold fusion. The latter comment implies that the Schwinger paper was accepted on the basis of his scientific stature (perhaps due to his Nobel prize?). Even so, the editor wanted to emphasize to the reader that "we can take no responsibility for the correctness of either the basic assumptions and the validity of the conclusions nor of the details of the calculations." These editorial comments are completely out of context in a scientific journal and serve to isolate these theoretical articles on cold fusion from the mainstream of scientific research.

One of the rather disturbing aspects of the polywater and cold fusion episodes was the great silence of most internationally respected scientists.

Although there were one or two experts speaking out strongly against both the polywater and cold fusion episodes, most experts chose to remain silent. Hildebrand voiced his very strong opposition to polywater research in a classic letter to *Science* (*ibid.*). Similarly, N.P. Samios, Director of the Brookhaven National Laboratory, wrote a negative article about cold fusion in the *New York Times* (September 24, 1989). On the other hand, prominent theoretical physicists, e.g. Edward Teller and Julian Schwinger, publicly supported cold fusion. Based on Fleischmann and Pons' most unusual claim, one would have expected more strong public statements against cold fusion. David Lindley in an editorial in *Nature* [**344** 376 (1990)] had some interesting comments on the response of most experts to cold fusion. He stated:

> [O]ne of the reasons that Pons and Fleischmann prospered early on was that few people were willing to stand up and say why they thought cold fusion was nonsense . . . [A]fter the March 23 press announcement, the response of most experts consulted by science reporters was academically correct but journalistically weak. It's a very interesting idea, was the gist of the scientific community's opinion, and we really can't say what we think of the experiments until we've seen more details, or tried it for ourselves, or consulted our colleagues in the chemistry department. This measured skepticism, contrasted with the unhesitant declaration coming from Pons and Fleischmann, sounded like academic nose-holding, as if physicists knew they had been beaten but could not bring themselves to admit it . . . Perhaps science has become too polite.

Although a number of fusion experts immediately proclaimed after Utah's appalling press conference that Fleischmann and Pons' startling findings were most likely a result of experimental errors, most when speaking to news reporters requested that their names be withheld. One expert who allowed his name to be used was Robert L. McCrory, director of the University of Rochester's Laboratory for Laser Energetics. He stated emphatically that "if I know anything about physics, [Fleischmann and Pons results] are wrong." He went on to say that getting four watts of heat energy from one watt input would require the emission of "a trillion neutrons per second – and in one hour they'd have a lethal dose" (*Business Week*, April 10, 1989).

In the summer of 1973, Deryagin and Churaev published a very short communication in *Nature* [**244** 430 (1973)] entitled "Nature of 'Anomalous water'." In this article Deryagin retracted ten years of research effort based on the application of more sensitive analytical techniques to test for impurities. "We have established that there are no condensates both free of impurity atoms and simultaneously exhibiting anomalous properties. Consequently, these properties should be attributed to impurities rather than to the existence of polymeric water molecules." At the time Deryagin wrote the above short communication admitting that polywater was a nondiscovery, an artifact caused by impurities, he was already over seventy years old. It was surprising then that his name should surface again, more than a decade later, in connection with the second major scientific episode of the century. In his laboratory in Moscow, the Institute of Physical Chemistry of the Academy of Sciences of the USSR, studies were conducted on the possibility of nuclear reactions occurring during the fracture of solids. Some of the reported results of these experiments and attempts to reproduce them, as well as their relationship to experiments of the type reported by Jones *et al.* are discussed elsewhere (Chapter VIII).

XII

Pathological Science

During the first week of April, 1989 I was in Berlin attending a conference commemorating the fiftieth anniversary of the discovery of nuclear fission by Otto Hahn and Fritz Strassman, who had carried out their historic research at the Kaiser Wilhelm Institute in Berlin. Since the University of Utah's press conference was only slightly more than a week old the animated discussions outside the lecture hall among this large group of nuclear scientists were confined largely to the University of Utah's reported room temperature nuclear fusion in a jar. Among the conferees was John A. Wheeler, distinguished nuclear theorist from Princeton University, who with Niels Bohr had published the seminal paper on the theory of nuclear fission in a 1939 issue of *The Physical Review*. On a bus ride from our downtown hotel to the Hahn-Meitner-Institute on the outskirts of Berlin, Wheeler expressed his views to me about cold nuclear fusion by comparing the University of Utah episode with René Blondlot's discovery of N rays. As will be described below, the N ray affair is one of the most remarkable known cases of self-deception in science which affected many French scientists in the early 1900s.

In 1903 Blondlot, a leading French physicist and member of the French Academy of Sciences, announced he had discovered a new kind of rays, which he named N rays, after the University of Nancy, where he did his research. Following the fundamental discovery of X rays by Roentgen in 1895, Blondlot in experimenting with X ray sources claimed he had found a new kind of emanation. These strange new rays could penetrate inches of aluminum but were stopped by thin foils of iron. The general properties of N rays were elusive at best. When N rays impinged on an object it was claimed that there was a slight increase in brightness. Blondlot admitted, however, that a great deal of skill was needed to see the effect of these rays. This did not keep a number of other physicists from reporting and extending Blondlot's findings. A large number of papers were published, many arguing that N rays ought to be important because X rays were considered to be one of the most important types of radiation known.

N rays do not exist. The many scientists who reported seeing them were the victims of self-deception. One of the interesting aspects of this episode is the large number of working scientists who were taken in. When the renowned American physicist R.W. Wood heard about these claims, he went to France to visit Blondlot's laboratory and to observe his experiments first hand. Blondlot at that time was using a prism to separate the different components of N rays. In a dark room Blondlot was demonstrating to Wood that he could measure three or four different refractive indices, each to two or three significant figures. The amazing feature of these experiments, as asserted by Blondlot, was that they were accurately repeatable. Wood listened and observed for a period of time, noticing that Blondlot was measuring the position of the beam of N rays to within a tenth of a millimeter. On asking how it was possible to detect the N rays with such great precision, Blondlot replied (as recorded in *Physics Today*, October, 1989):

> That's one of the fascinating things about N rays. They don't follow the ordinary laws of science that you ordinarily think of. You have to consider these things all by themselves. They are very interesting but you have to discover the laws that govern them.

By now Wood had correctly surmised that something peculiar was happening. In the darkened room Wood surreptitiously removed the prism and placed it in his pocket. He then politely asked Blondlot to repeat some of the previous measurements, which he was happy to do. With the centerpiece of the experiment missing, Blondlot obtained exactly the same results. Wood wrote a devastating account of his observations, which was published in *Nature*, showing Blondlot's N rays resulted from self deception.

Wood's exposé finished N rays outside France, but French scientists continued to support Blondlot for some years. Science is supposed to transcend national and other boundaries, but Wood's critique of Blondlot's N rays did not. How was it possible for French scientists to continue to support Blondlot and research in N rays after Wood's critique? In this case it appeared to be due to an uncritical acceptance of N rays driven by national pride. French scientists at this time felt their international reputation in science was weak, at least relative to German science. The discovery of N rays served to bolster their self image at a critical time. Members of the French Academy's Leconte prize committee, which included Henri Poincaré, chose Blondlot for its 1904 Leconte prize. This occurred in spite of the fact that among the candidates was Pierre Curie, who had shared the Nobel prize in 1903 for the discovery of radioactivity. By choosing Blondlot over Pierre Curie for the prestigious Leconte prize after Wood's exposé and the strong criticism of N rays from abroad, the prize committee demonstrated its zealous support for Blondlot and N rays rather than acknowledge openly that the whole N ray episode was a classic case of self-deception.

In 1953 Irving Langmuir, Nobel laureate in chemistry, gave a colloquium at General Electric's Knolls Atomic Power Laboratory on pathological science, a subject he had explored for some time. Langmuir's lecture was a colorful account of some of the pitfalls into which scientists may stumble. Robert N. Hall, a former colleague at General Electric, made an internal report of Langmuir's talk for the company from a disk recording found among the Langmuir papers in the Library of Congress. I had a copy of this report in my file which I retrieved and read again with renewed interest on returning from Berlin in early April. *Physics Today* (October, 1989, p. 36) has done a service to science by making Langmuir's remarks available to a wide audience.

Langmuir defined pathological science as "the science of things that aren't so." He analyzed the symptoms seen in studies of N rays and several other elusive phenomena and suggested six criteria of pathological science (*ibid.* p. 44).

(1) The maximum effect that is observed is produced by a causative agent of barely detectable intensity, and the magnitude of the effect is substantially independent of the intensity of the cause.

(2) The effect is of a magnitude that remains close to the limit of detectability or, many measurements are necessary because of the very low statistical significance of the results.

(3) There are claims of great accuracy.

(4) Fantastic theories contrary to experience are suggested.

(5) Criticisms are met by *ad hoc* excuses thought up on the spur of the moment.

(6) The ratio of supporters to critics rises to somewhere near 50% and then falls gradually to oblivion.

In addition to N rays described previously, how many other far-out new ideas in the twentieth century have been relegated to the scientific graveyard of pathological science? There are several. One of the other examples discussed by Langmuir is the case of the Allison effect. Dr. Fred Allison, a professor at the Alabama Polytechnic Institute, invented a technique for detecting elements and isotopes. The Allison effect was utilized to 'discover' several new elements such as alabamine and virginiam which were listed at the time in the Discoveries of the Year. Hundreds of papers were published on this strange effect in America's most prestigious journals. Even Wendell Latimer, well known chemist at Berkeley, was taken in and used the Allison effect to 'discover' the heavy isotope of hydrogen of atomic weight three, publishing his results in the *Physical Review*. All results based on the Allison effect were not reproducible and turned out to be erroneous. The experiments utilizing the Allison effect fit the general characteristics of pathological science. The number of publications on the Allison effect reached its peak in the 1930s and then faded into oblivion. Other examples of pathological science examined by Langmuir were extrasensory perception and flying saucers.

Before examining cold fusion by Langmuir's criteria, three other examples of pathological science in more recent history (post-Langmuir) are described briefly, one from the field of chemistry, one from high energy physics and one from biology. Polywater, the example from chemistry, attracted major attention in the late 1960s and early 1970s as described in Chapter XI. Polywater like cold fusion was involved with water, one of the most abundant substances on earth. Felix Frank (*Polywater*, MIT Press, 1981) stated the case for the polywater episode as belonging to pathological science.

> Measured against Langmuir's six criteria, polywater certainly exhibits the symptoms of pathological science. Never more than a few micrograms were available for any analytical or physical measurements, so that the experimental techniques had always to be pushed to the ultimate limits of instrumental sensitivity. The liquid could never be made to condense in all the capillary tubes of any batch, but only in about 30 to 40 percent. The very existence of the newly discovered substance rested on the premise that some completely new and unknown type of chemical bond was involved. Doubts relating to the presence of impurities were brushed aside by statements that in the believer's laboratory all due care was taken to exclude impurities. Finally, the ratio of supporters to critics also closely followed the course predicted by Langmuir.

As was the case in the N-ray episode, other factors beside irrational and aberrant behavior on the part of the scientists may have also played a role in the polywater episode. As already mentioned, personal competitive drives of the scientists, and institutional and national loyalties sometimes play a role in distorting science. In the case of polywater, self-deception was a major ingredient of the whole affair and in that sense it deserves the classification of pathological science.

Physicists studying projectile fragments produced by the interactions of relativistic heavy ions in the cosmic radiation and from high-energy accelerators on various targets reported new particles dubbed anomalons. These secondary projectile fragments, reported to have anomalously short mean free paths in relativistic heavy-ion collisions, attracted considerable interest in recent years. Although the original evidence for this anomalous behavior of secondary projectile fragments came from cosmic ray research as early as 1954, activity in this area markedly increased in the 1980s with the availability of accelerator-produced relativistic projectiles.[27] Experi-

---

[27] Several of the papers on anomalons supposedly made by accelerator-produced relativistic projectiles have been due to E.M. Friedlander and collaborators, for

menters admit that the reported short mean free path of the secondary fragments cannot be explained within the framework of conventional physics. Although at this time no final judgement can be made, anomalons have several of the characteristics of pathological science. Some experiments show anomalous behavior, others do not. Problems of limited statistics, systematic errors and reproducibility have kept these mysterious observations from being generally accepted. Furthermore, activity in this field is rapidly diminishing and may be in the final Langmuir stage of falling gradually to oblivion.

On June 30, 1988, a group of biological scientists from France, Israel and Canada published a joint paper in *Nature* [**333**, 816 (1988)] claiming that an aqueous solution of an antibody retains its ability to evoke a biological response even when diluted by astronomically large factors. Benveniste and his colleagues reported calculations that showed less than one molecule of antibody was present in their assay when anti-IGE antiserum was diluted by a factor of 10^{14} (corresponding to 2.2×10^{-20} molar). In their remarkable experiment, they reported detecting biological activity down to a dilution of 10^{120}. At such dilutions[28] the chance of finding a single molecule of antibody is infinitesimally small! Hence, most of the experiments with antibody solution reported by Benveniste *et al.* were carried out in the absence of any antibody molecule.

There is no objective explanation for the claimed observations of Benveniste and collaborators since there is no physical basis for biological activity in solutions without antibody molecules. Although *Nature* published the strange results of Benveniste *et al.*, it was on the condition that a team could be sent later to Benveniste's laboratory to investigate the procedures used in the high-dilution experiments. This team included Walter W. Stewart (who is known for his studies of misconduct in science), James Randi (the professional magician known as "The Amazing Randi"; also a MacArthur Foundation fellow) and John Maddox (the Editor of *Nature* with a background in theoretical physics). On July 28, 1988, *Nature* published [**334** 287 (1988)] the findings of this team in an article entitled "High-Dilution Experiments a Delusion." This team of three concluded:

> The remarkable claims made by Dr. Jacques Benveniste and his associates are based chiefly on an extensive series of experiments which are statistically ill-controlled, from which no substantial effort has been made to exclude systematic error, including observer bias, and whose interpretation has been clouded by the exclusion of measurements in conflict with

example, *Phys. Rev. Lett.* **45** 1084 (1980); **55** 1176 (1985); *Phys. Rev.* **C27** 1489 (1983), **C38** 1658 (1988).

[28] Paul J. Karol pointed out (*C&EN*, September 5, 1988, p. 38) that concentrations of 10^{-120} molar require volumes corresponding to many billions of universes.

the claim that anti-IgE at 'high dilution' will degranulate basophils. The phenomenon described is not reproducible in the ordinary meaning of that word.

There are many similarities with cold fusion; lack of control experiments, statistical uncertainties, irreproducibility and the public description as a "simple experiment."

Benveniste and his supporters continue to believe that their "high-dilution" experiments are correct. This view or theory of biological activity at extreme dilution has been held for over a century and a half by those working in the field of homeopathic medicine. The search for verification of this theory will go on because a sizeable fraction of French physicians prescribe homeopathic medicines. Much has been written about the "Benveniste effect", however, there is no reproducible experimental evidence to support it. The effect clearly has the characteristics of pathological science, namely, "the science of things that aren't so." *Nature* has been criticized for publication of the paper by Benveniste *et al.* and for their team's investigation of Benveniste's claims. This sequence of events no doubt influenced *Nature*'s handling of many of the more bizarre contributions on cold fusion.

Cold fusion as an energy source will likely go down in history as a colossal aberration of scientific progress. Had Langmuir lived to experience the bizarre episodes of cold fusion and polywater, he would no doubt have pursued his avocation on the study of pathological science even more vigorously. It would have been most interesting to have heard his analysis and comments on these two major events involving water. Both of these strange episodes in science had the potential of imposing major changes on society. Room temperature fusion in a jar was a fantasy of almost everyone, especially of environmentalists and energy specialists. The month of April, 1989 was a high point when perhaps several hundred million people from around the world had heard of Fleischmann, Pons and cold fusion and were led to believe in the dream of a new essentially limitless source of cheap, safe and clean energy. The ensuing months raised havoc with their dream. From the beginning there was a terrible discrepancy between the intensities of the claimed excess heat and the fusion products. Then it soon became known that many groups could not replicate the early positive claims. Even so, there are still scientists who believe very strongly in cold fusion.

Judged on the basis of Langmuir's criteria, cold fusion in lattices displays some of the symptoms of pathological science. First some comments about the magnitude of the effect and its statistical significance. A major fraction of the experimenters attempting to replicate cold fusion produced neither excess heat nor fusion products. Others, however, reported excess heat and either no fusion products or fusion products at a level orders of magnitude below that implied by the reported excess heat. Of particular importance is the fact that all of the experiments reporting positive results were plagued by the lack of reproducibility and many by internal inconsistencies. Claims of fusion products at very low levels near background were inconsistent with other experiments that found no fusion products and placed upper limits on the fusion probability well below the claimed positive results. Based on the many negative results on calorimetry and fusion products and the marginal statistical significance of reported positive results, most scientists agree with the early conclusions of our DOE/ERAB panel that the evidence for a new nuclear process called cold fusion is not persuasive. Most faced up to the totality of evidence soon and rejected the proposition of Fleischmann and Pons that an unknown nuclear process (or processes) could bridge the irreconcilable discrepancy between the claimed intensities of the excess heat and fusion products. Some, however, still believe in cold fusion and may continue to do so. This is probably not too surprising based on previous history. After Blondlot's N rays were exposed as illusionary, he received the Leconte prize and papers continued to be published on the subject for some time. Five months after our DOE/ERAB negative report, approximately 250 believers in cold fusion met in Salt Lake City to rehash room temperature fusion. The audience's faith in cold fusion was demonstrated following Fleischmann's inspirational summary talk at the First Annual Conference on Cold Fusion when they rose to their feet and applauded vigorously.

Cold fusion proponents have produced fantastic theories contradicting both experimental and theoretical evidence in nuclear physics. In this respect cold fusion fits squarely into Langmuir's pathological science. Some examples follow. Fleischmann and Pons postulated that deuterium is compressed electrochemically in metallic lattices to astronomical pressures ("such high hydrostatic pressures are naturally not achievable on earth") squeezing the deuterium close enough to fuse, whereas all evidence indicates that the D–D separation distance in the palladium lattice is larger than the known separation distance of 0.074 nanometers in the D_2 molecule. As stated in the previous chapter, one of the surprising aspects of the cold fusion saga has been the large number (somewhere between 50 and 100) of theoretical papers reporting 'positive' results. Most of these papers start from the assumption that cold fusion is experimentally established and then proceed to 'explain' the claimed experimental results by invoking one or more marvels or miracles. Such features of these 'theories' are contrary to experience and are characteristic criteria of pathological science. One responsibility of theorists is to guide experimentalists. Fabricating theories 'to explain' erroneous experimental data is counter-productive.

Proponents of cold fusion had no hesitation in arbitrarily changing the well known branching ratios of the D+D reaction channels when seeking to "explain" unconfirmed data which could not be reproduced in most laboratories. Advocates even suggested that high-energy nuclear gamma rays could be concealed by a miraculous process in which the entire reaction energy was transferred into the solid lattice! The 'theory' of Walling and Simons (see Chapter III) of the University of Utah is the ultimate example of Langmuir's fourth criteria of pathological science. It is ironic that following this contribution Simons, one of the co-inventors of cold fusion, was named the first Henry Eyring Professor of Chemistry at the University of Utah (*Chemical and Engineering News*, August 28, 1989). Recall that Blondlot received the Leconte prize after Wood's exposé of N rays.

The cold fusion claims have been riddled with inconsistencies. The many scientific groups that could neither produce excess heat nor produce fusion products in cold fusion experiments were deluged with excuses for their failure from the believers. These excuses ranged the full gamut from "it's hard for them to see the effects because they are doing it wrong" to the caustic statement that negative results are of little significance because "they do not require any special skill or expertise in electrochemistry." Bockris and others claimed the effect does not turn on for many weeks of electrolysis and this explained to the believers the negative results from 'Eastern establishment' universities who ran for shorter times. However, this explanation was inconsistent with the results of the BARC groups who had success the first day of electrolysis. Some believers claimed high currents were required while the original Fleischmann and Pons' paper reported excess heat with currents as low as 8 and 64 milliamperes per square centimeter of palladium. When Pons was faced with the evidence that his cells produced no fusion products during a critical five-week period in May and June, 1989, he retorted that his cells were not producing excess heat during this period, even though this was inconsistent with a statement he made on August 16. Some proponents of cold fusion claimed negative results were due to the D/Pd atom ratio being below unity, while at the same time many of the believers were reporting success with ratios considerably below one. Measured against Langmuir's fifth criteria, cold fusion certainly exhibits this symptom of pathological science.

The ratio of the worldwide positive results on cold fusion to negative results[29] peaked at approximately 50% some five to six weeks after the March 23, 1989 press conference, qualitatively in agreement with Langmuir's sixth criteria. Morrison[29] has shown, however, that the positive to negative ratio is very sensitive to the geographical region under consideration. Since many of the negative results have not been published and most groups getting negative results soon returned to their own fields of research, most present activity in cold fusion is limited to believers. Fritz Will, the

[29] Douglas Morrison, *Cold Fusion News* No. 20, October 20, 1989.

Director of the National Cold Fusion Institute, has prepared a list of some 100 groups that claim to have produced at least one of the phenomena associated with cold fusion, i.e. either heat or one of the fusion products. This list, however, doesn't contain a single entry where the claimed heat is accompanied by a commensurate number of fusion products. Most of the entries in the list do not meet the standards of the peer-review process for journal publication. It is unfortunate that so much emphasis has been placed by the proponents on the length of this list of claims rather than on the confirmation and reproducibility of a single claim. When discussing the quantity of evidence, one must remember that in killing polywater Deryagin wrote off ten years of work in a published two-sentence retraction.

Morrison has extended Langmuir's study of wrong results in science and has added a number of additional characteristics over the last fifteen years to help in the identification of pathological science. In his detailed analysis, Morrison finds that cold fusion has a large fraction of his characteristics which define pathological science. He concludes that cold fusion is best explained as an example of pathological science (Lecture at the World Hydrogen Energy Conference, July 24, 1990, Honolulu; CERN preprint, CERN/PPE 90–159).

Believers in cold fusion will continue, at least for some period of time, to report claims of success. This pattern of continuing claims has been well established for other areas of pathological science such as N rays and polywater. What is surprising, however, is that these far-out claims are still reported with a positive flavor by leading newspapers (*Wall Street Journal*, April 8, 1991 and *New York Times*, April 14, 1991) before the most elementary confirmation checks have been made. Both of these newspaper articles describe recent experiments by a group of chemists at the Naval Weapons Center in China Lake, California, and give some scientific respectability to the chemists claim of producing ^4He in an electrochemical cell (see Chapter VIII). The human desire for a way to turn hydrogen, the most abundant fuel in the universe, into a limitless supply of clean, cheap fuel is so strong that people want to believe that there is something scientific about cold fusion.

One of the latest episodes to hit the press is a release by Randall L. Mills announcing a new method for generating enormous amounts of energy during the electrolysis of water, not by a nuclear process as claimed by Fleischmann and Pons, but by a *chemical* process (a power output of 37 times the power input!). In Mills' process the source of heat is the "electrocatalytically induced reaction whereby hydrogen atoms undergo transitions to quantized energy levels of lower energy than the conventional ground state"! Aside from the claimed excess power gain of hundreds of times that of Fleischmann and Pons, one of the most striking things about this unconventional claim is that it will be published in the *Journal of Fusion Technology*.

More than two years have elapsed since Fleischmann and Pons made their startling claim to have observed nuclear fusion at room temperature. At the present time the ratio of experimental papers reporting "positive" to "negative" results is much larger than that during the early months. Proponents of cold fusion have attached particular importance to this trend suggesting that the experimental evidence for cold fusion has improved significantly over the last two years. A more realistic interpretation of the present larger ratio acknowledges the sharp decline in the number of negative papers due to the fact that many of the earlier experimenters failed to produce any evidence for cold fusion and returned to their own special fields of research. Hence, the present experimental and theoretical work in the field of cold fusion is limited mainly to believers who continue to report claims of positive results. The membership in this club of believers reporting positive results has remained essentially unchanged over the last year.

Two recent reviews of cold fusion illustrate that the cold fusion phenomenon lives on, at least in the minds of believers. One review by M. Srinivasan of BARC is entitled "Nuclear Fusion in an Atomic Lattice: An Update on the International Status of Cold Fusion Research" is published in *Current Science* (April 25, 1991). The other review by Edmund Storms of the Los Alamos National Laboratory is entitled "Review of Experimental Observations About the Cold Fusion Effect" is to be published in *Fusion Technology* (1991). These two reviews serve a useful purpose by compiling 174 and 366 references, respectively, to the cold-fusion literature. However, neither of these compilations include all results nor do they critically compare the data presented. A more extensive Bibliography with currently over seven hundred references to cold fusion has been compiled by Dieter Britz of Aarhus University (this Bibliography which usually gives a concise summary of each entry, is distributed electronically; a copy can be obtained through the Cornell Cold Fusion Archives).

Srinivasan included an interesting explanatory letter with his review where he states:

[I]n my judgement (there is) very convincing evidence for the authenticity of the (cold fusion) phenomenon and its nuclear origin . . . I am amazed to see the intense anti-cold fusion propaganda launched by a handful of influential people at the international level.

Srinivasan's belief that the negative point of view on cold fusion which has developed over the last two years is due to a handful of influential skeptics is strong evidence of self-deception, a characteristic of pathological science. In sharp contrast to Srinivasan's view on cold fusion, most scientists have written off cold fusion at least at the level claimed by Srinivasan, where the process produces watts of excess power due to a nuclear process. By Srinivisan's own testimony only a small band of "about 600 scientists" the world over continue to study cold fusion (this number seems rather large). If the scientific community at large believed that cold fusion was a potential source of abundant, cheap and clean energy can you imagine most scientists the world over abandoning cold fusion research? Hence, it is entirely unrealistic to place the blame for the current "back-to-the-wall" predicament of the cold fusion proponents onto a highly vocal small group of dissenters. For example, Srinivisan's attempt to dismiss Frank Close's negative book [*Too Hot to Handle*, Princeton University Press, (1991)] as the work of a hot fusion advocate is based on the fact that this book was published by Princeton University Press (and it so happens that Princeton University operates the hot-fusion facility known as Tokamak)! The same malicious and false accusation of hot fusion interest was leveled by several believers against members of our DOE/ERAB panel in an attempt to discredit our study of cold fusion. For example, Eugene Mallove, one of the more ardent believers in cold fusion, inferred that my negative stand on cold fusion claims is due to the large inertial confinement fusion project at the University of Rochester. Had he taken the trouble to check with anyone at the University of Rochester, he would have learned that I have no association with this project. To assume that skeptics of cold fusion are hot fusion proponents is simply faulty logic and a desperate move to discredit those who find that the evidence for cold fusion is not compelling.

Storms' review is equally optimistic to that of Srinivisan. He concludes,

The number and variety of careful experimental measurements of heat, tritium, neutron and helium production strongly support the occurrence of nuclear reactions in a metal lattice near room temperature as proposed by Pons, Fleischmann and Jones.

What is the meaning of this sentence where the high-level power claims of Pons and Fleischmann are lumped together in a single sentence with Jones' claim of low-level neutrons? Since these two claims differ by a factor of more than a thousand billion, what am I to believe is strongly supported? In a letter to me accompanying his manuscript, Storms stated:

Since the ERAB Cold Fusion Panel report was written, considerable information has been published that strongly supports the claims of Pons, Fleischmann and Jones. In addition, a number of new aspects to the phenomenon have been discovered. In view of this new information, the great importance [of] this discovery, and the generally negative evaluation given in the early ERAB report, I would like to suggest that a , second look be made by the ERAB panel.

If compelling evidence for cold fusion were discovered in the interval following our panel's final report, Storms' suggestion is a very reasonable one. A thorough study of these two reviews along with the original papers that are referenced therein shows, at least in my opinion, that there is still no persuasive experimental evidence for the phenomenon known as cold fusion.

How is it possible for different people to examine the same experimental claims and reach opposite conclusions? First of all, both believers and skeptics acknowledge that there are a large number of experimental claims (it is immaterial whether the number is 50, 100 or some other number, depending on how the independent claims are counted) supporting some aspect of cold fusion. On the basis of the sheer number of positive claims, it is tempting to conclude, as many believers have, that there must be some truth to cold fusion. Numbers of claims alone, however, are not definitive in science. Recall that several hundred papers were published in support of both N rays and polywater, two classic cases of pathological science. It is imperative, therefore, to dissect each claim and examine its validity. This close in-depth analysis of individual claims often leads skeptics and believers to diametrically opposite conclusions.

The unmistakable signature for the occurrence of nuclear fusion of deuterons (D+D) is the production of fusion products (see Table 1 on p. 109). Heat, if due to nuclear fusion, must be accompanied by a commensurate amount of fusion products. Once one abandons this equality, one has left science as it is normally practiced. One careful experiment showing an equality between heat and fusion products would settle the issue. However, two years have elapsed and there is still *not a single claim* where the reported heat is accompanied by a commensurate amount[30] of fusion products! In fact the two quantities differ by many orders of magnitude. This acknowledged large inequality, however, hasn't convinced the believers that the heat cannot be nuclear in origin. Some believers even use the very low-level neutron claims of Jones *et al.* as proof that the reported heat is due to a nuclear fusion process, even though the two claims may differ by up to thirteen orders of magnitude. Based on these very different interpretations of the same reported observations, it is understandable how believers and skeptics, respectively, conclude that the claimed excess heat *is* and *is not* due to nuclear fusion. Fusion heat without fusion products is a high-order miracle which skeptics are not willing to swallow based on the presently

---

[30] The recent rehashed claim of Bush *et al.* [*J. of Electroanalytical Chemistry* **304** 271 (1991)] of a commensurate amount of ^4He in the effluent gases from an electrolytic cell is unconfirmed. First, no evidence for the commensurate intensity of the 23.8–MeV gamma rays was presented. Secondly, no ^3He was observed as required. Its absence requires a miraculous alteration of conventional low-energy D+D fusion.

reported fragmentary and sometimes contradictory experimental evidence for both heat and particles.

The reviews of Srinivasan and Storms conclude that the present observational claims of producing neutrons, protons and helium from cold fusion experiments employing electrochemical and high-pressure gas cells strongly support the existence of room temperature fusion in a metal lattice. I disagree completely with this conclusion. My arguments for concluding that no compelling experimental evidence exists for the production of fusion products in metals at room temperature are given in Chapter VIII. Individual scientists will have to examine the primary literature themselves in order to make their own conclusion about the validity of the various claims. The most careful searches for fusion products have been for neutrons. Although fusion neutrons at a very low level cannot be categorically ruled out, the present evidence for room temperature fusion neutrons is also not at all convincing.

In Section 22 of Srinivasan's review, he lists nine "puzzles" that summarize the experimental claims of cold fusion advocates and that will need to be explained by any cold fusion theory. Each of these several "puzzles" in the nuclear area contradict conventional nuclear physics and require acceptance of a series of "miracles" (see Chapter VIII). Srinivasan does a service to the cold fusion controversy by admitting that the evidence for cold fusion requires acceptance of this series of "puzzles", like, for example, the anomalous T/n ratio of about 10^8 rather than one. Believers and skeptics look at this type of experimental evidence very differently, especially when only two laboratories in the world have claimed high levels of tritium in experiments which are not reproducible. When miracle after miracle are required to accept the claims of heat from room-temperature fusion which are contrary to experience, one is likely dealing with pathological science.

The Second Annual Conference on Cold Fusion was held in Como, Italy on June 29–July 4, 1991. On the International Advisory Committee were Bockris, Fleischmann, Menlove, Pons, Preparata, Scaramuzzi, Srinivasan and Will, all familiar names as strong believers in cold fusion. One name missing from the International Advisory Committee is Steven E. Jones. Although invited to participate, Jones declined for two reasons. First, he disagreed with the premise that the claimed "excess heat" was due to nuclear reactions. Jones stated that his own research showed that the nuclear products were many orders of magnitude too small to correlate with "excess heat", refuting the notion of Fleischmann and Pons that the excess power is due to fusion. Secondly, at least one member of the committee had used threats of legal action against fellow scientists and colleagues. Jones stated that he could not in good conscience join with such a member on the Advisory Committee.

Some of the milestones in the cold fusion saga leading up to the Second Annual Conference are listed in Appendix III. There is now an open rift in the two camps of the cold fusion community. Jones refused to attend this conference because he wants to disassociate his claim of very low particle intensities from the claim of excess heat due to room temperature nuclear fusion. And well he should. The thirteen orders of magnitude difference between watts of excess power and his claim of neutrons are irreconcilable to Jones and most other nuclear physicists. Jones has taken this position from the beginning.

How would the history of cold fusion have developed if Fleischmann and Pons had not been so impressed on February 23, 1989 with the BYU evidence for neutrons? One possible scenario is that the University of Utah group would have continued at their own pace and taken the extra eighteen months that Fleischmann is quoted to have wanted. During this time, collaborative arrangements would have been made with nuclear physicists to search for fusion products with state-of-the-art counting equipment. On finding the upper limits of fusion product intensities to be many orders of magnitude smaller than their reported amounts of 'excess heat', they would have designed a second generation of calorimeters looking for some non-nuclear explanation of their results. In the meantime, the BYU group might have proceeded to publish their neutron results, borderline as they were. These very low intensity neutron results would have appeared in the literature with little fanfare and with few taking notice outside of a few specialists. After all, Deryagin and others in the Soviet Union had published a number of such papers over the last several years in which they reported low levels of neutrons. However, it was Fleischmann and Pons' decision to go public with their far-out claim of watts of power from room-temperature fusion that propelled the BYU experiments into the limelight.

Fleischmann and Pons' gamble to go public with their announcement of watts of 'excess power' from nuclear fusion at room temperature, before they had any solid evidence of fusion products, is the scientific fiasco of the century. The chasm between Fleischmann and Pons' claim of 'excess heat' and the upper limit on the intensities of the associated fusion products continues to be many orders of magnitude! Room temperature nuclear fusion without commensurate amounts of fusion products is a delusion and qualifies as pathological science defined as "the science of things that aren't so." Only time will tell whether there are processes such as, for example, 'fracto-fusion' that can account for very low levels of fusion products. The present evidence is not persuasive.

XIII

Lessons

Some would dismiss polywater and cold fusion[31] as local aberrations in science. Such a view is too simplistic. Each of these two unusual phenomena engaged a large number of scientists cutting across several different disciplines. Eventually both phenomena were woven deeply into the fabric of science. The study of cold fusion, for example, involved chemists, materials scientists and physicists including Nobel laureates. Many tens of millions of dollars have been spent on cold fusion research trying to replicate a far-out idea. It seems wise, therefore, in this final chapter to discuss some general issues of importance to the advancement of science. The topics selected are neither all-inclusive nor unique to cold fusion. Both advocates and opponents of cold fusion agree that useful lessons can be learned from the way this controversy developed. Some of the hypotheses, procedures, decisions and actions associated with the cold fusion saga will be examined. A range of issues come to mind from the fundamental treatment of data, proper experimental controls and judging of hypotheses to the management of research at all levels. More specifically, how did cold fusion germinate and what fueled the whole episode while reaching such gigantic proportions? Was the dream of limitless, clean and cheap energy so powerful that it bred error and self deception? In addition to the researchers themselves, referees and editors also have essential and important responsibilities in the communication of scientific results. The failure of one or more of these groups to function in a responsible way is highly detrimental to the scientific process.[32]

---

[31] Cold fusion here refers to room temperature nuclear fusion producing watts of power as claimed by Fleischmann and Pons.

[32] In writing some sections of this chapter I have benefitted from the booklet entitled *On Being a Scientist* published recently by the National Academy Press, Washington, D.C. (1989). I highly recommend that all students of science read this booklet.

Handling Far-Out Ideas and Claims

By labeling cold fusion a far-out idea, some may accuse me of the understatement of the year. Fleischmann and Pons' far-out idea was to use electrochemical techniques to fuse deuterium at room temperature! Chemistry was used in order to attempt to increase the rate of a nuclear reaction by over 50 orders of magnitude (10^{50}). They postulated that by loading deuterium into palladium by electrolysis of D_2O sufficiently large pressures would be developed to squeeze deuterium nuclei close enough to fuse. How should scientists handle such far-out ideas and claims? Presentation at a small informal meeting of colleagues is the normal procedure for testing and evaluating far-out ideas and/or experimental claims which contradict well-established experimental and theoretical results. Only when colleagues have had an opportunity to ask hard questions and point out the obvious mistakes, inconsistencies and the evident checks needed to be made, and their suggestions are addressed, does one think about going public with far-out ideas and claims. Local presentations to multidisciplinary audiences may be of special benefit when the idea or claim is partly outside ones own discipline. The next step would normally be to present the results to experts off campus and eventually submit to the peer-review process.

The first instinct of an experimental scientist, when confronted with an unexpected far-out result, is to try to make it go away. Every effort has to be made to track down all possible conventional explanations. There is no simple recipe to follow when attempting to kill an unexpected result. The experimental apparatus and analysis procedure may have to be revamped to insure that the same results are obtained with many different kinds of equipment and in different circumstances. Fleischmann and Pons were unable to convince others that they had made a genuine attempt to kill their claims. For example, their neutron measurements were done so poorly that within a few weeks others had shown that Fleischmann and Pons' reported neutron results were artifacts. As a consequence, scientists were skeptical of all their claims. Furthermore, although Fleischmann and Pons recognized the enormous disparity between their claimed amounts of excess heat and fusion products, they never adequately addressed this contradictory evidence. The simple remark in their first paper stating that the heat was due to an unknown nuclear process or processes did not sit very well with nuclear physicists. A true scientific phenomenon is invariant. Other experimenters working with different equipment and in very different circumstances must be able to repeat and confirm a far-out result if it is valid.

When advocates of cold fusion claimed to produce heat without fusion products or tritium without neutrons, one realized that some very basic principles were being challenged on which whole technologies have been

216

based. Such far-out claims lack scientific credibility and raise a series of questions about the methodology used in reaching them.

Judging Hypotheses

When new reported observations come into direct conflict with long-established experimental and theoretical results, individual responses range between two extreme reactions. One extreme reaction is to reject all previous results and the second extreme is to hold onto an incorrect experimental result or outdated theory for a time long after one or both have been discredited.

The history of science contains numerous examples where scientists have embraced a favorite but discredited idea for far too long a period. When convincing evidence is forthcoming, scientists must be willing in the face of new results to change their position. Some cold fusion proponents erred in the other extreme by modifying well-tested results in nuclear science when fragmentary but irreproducible cold fusion data were reported. Again the history of science contains examples where such premature rejection of valid science occurred before there was any experimental justification to do so.

Occasionally surprises do occur in science and one should be prepared for them. Just over fifty years ago when nuclear physics was in its infancy, the discovery of nuclear fission came as a complete surprise. Several months before Hahn and Strassman's discovery of fission, Fermi, one of the world's most distinguished nuclear physicists, misidentified fission products. He assigned the then unknown radioactive species to new transuranium nuclides rather than to fission fragments because he had not thought of the idea that a nucleus could split into two heavy nuclei. As a field matures, as nuclear physics has over the last half century, the probability of a surprise becomes ever so much less probable.

During the short lifetime of cold fusion, its advocates have required belief in miracle after miracle for one to swallow their discovery, suspending each time some existing knowledge in nuclear physics. The claimed fusion rates in solids were justified by creating heavy electrons, a mechanism that would have had to have escaped the attention of nuclear physicists for generations. The well known branching ratio of the D+D reaction was miraculously altered, in one case to completely suppress the dominant neutron and tritium branches and in another case to give tritium to neutron ratios of over a million, compared to the well known ratio of approximately one. The claimed absence of high energy gamma rays from the $D+D \rightarrow {}^4He+\gamma$ and $D+H \rightarrow {}^3He+\gamma$ reactions was accounted for by yet other miracles. The theory of Walling and Simons, described in Chapter III, required all of the above miracles. Such a pyramid of miracles is strong circumstantial evidence of pathology at work. Bockris and others have gone so far as to state that conventional nuclear physics is applicable only to gases. Hence, in solid metal lattices, these proponents of cold fusion assumed that all previous knowledge about the D+D reaction is invalid. This led Bockris to assign the claimed T/n ratio of 10^6 to 10^9 as one of the two fundamental characteristics of cold fusion (the other being irreproducibility). Fleischmann and Pons started their research program in cold fusion because they mistakenly believed that the pressures attained during the electrolysis of D_2O were so large that the deuterons were miraculously squeezed closely enough together for fusion to occur in palladium lattices.

Premature Publication

The rush to publish goes back to the seventeenth century when the Royal Society of London gave priority to the one who published first, rather than to the one who first made the discovery [*Science* **251** 260 (1991)]. Researchers have been competing to publish their results first ever since. One of the major worries about rapid publication is that it may be premature publication. In the rush to publish, scientists eliminate important experimental checks of their work and editors compromise the peer-review process, leading to incomplete or even incorrect results. Most journals wish to publish quickly, but also want to be accurate. One solution to publishing competing papers is to have a journal coordinate publication so that both papers appear simultaneously. This can best be done by the authors, but requires some give and take. As an example of coordinated publication, Jones claimed that an arrangement had been made with Pons and Fleischmann whereby both groups were to send their manuscripts simultaneously by Federal Express to *Nature* from the Salt Lake City airport on March 24, 1989. This, however, did not happen.

Journals sometimes compete with each other in publishing seminal papers from rival groups. Faster publication can sometimes be insured by telling the editor that a competing paper is due to be published in another journal. Editors have reacted to the pressure to publish quickly in various ways. On April 26, 1976 the *Physical Review Letters*, for example, instituted a policy that publication of experimental papers could be requested without peer review provided the request was seconded by the department chair or an equivalent administrator. There was a catch though. If this request was made, the journal would print a disclaimer on the first page of such articles informing readers that, at the author's request, it had not been peer reviewed.

On June 16, 1976 R.V. Gentry *et al.* submitted an article to the *Physical*

Review Letters entitled, "Evidence for primordial Super-heavy Elements." This article was published on July 5, 1976 (**37**, p. 11), 19 days after submission, with the disclaimer "Accepted without review at the request of Alexander Zucker under policy announced 26 April 1976." In this paper, Gentry *et al.* claimed to have seen L x-rays of superheavy elements such as Z=126 from monazite (a mineral that contains rare-earth elements and thorium) samples bombarded with low-energy protons. These protons were claimed to produce L electron vacancies in the primordial element Z=126 leading to specific L x-rays. They stated that the evidence for element Z=126 was very solid because they had seen the characteristic L x-rays in five out of six samples. Following immediately was a theoretical paper explaining the stability and long lifetime of element Z=126 [F. Petrovich *et al.*, *Phys. Rev. Lett.* **37** 558 (1976)]. On July 7, 1976, two days after the publication of the Gentry *et al.* paper, another group submitted a Comment [J.D. Fox *et al.* *Phys. Rev. Lett.* **37** 629 (1976)] showing the previously claimed L x-ray line of element 126 was instead a gamma ray transition in ^{140}Pr excited by the ^{140}Ce (p,n) ^{140}Pr reaction, ^{140}Ce being a major constituent of monazite. Here is an example of very rapid publication where the experimental result was totally incorrect, and the biggest worry associated with the rush to publish came true. Fortunately, most authors do not want the stigma associated with the tag line, and only a very limited number of papers have been published by the above mechanism where peer review is bypassed by request.

The cold fusion paper of Fleischmann and Pons was published in the *Journal of Electroanalytical Chemistry* in just four weeks. This was a classical case of premature publication. An enormous amount of errata followed, including even the name of M. Hawkins, the third author omitted from the original paper! The nuclear measurements were done very sloppily and under normal circumstances, these measurements should have been repeated with more sophisticated experimental equipment before publication. Instead it was left for others to prove that the Fleischmann and Pons neutron results were artifacts. The more important the reported result the more imperative it is for investigators to take the responsibility to examine and check their data as well as to assess the reasonableness of the resulting conclusions reached. Journal editors also have a responsibility to screen out substandard papers. In the case of cold fusion, some manuscripts were accepted without peer review and others were accepted with inadequate peer review. Authors of most of the positive papers neither ran sufficient control experiments nor properly assessed the experimental errors in their primary measurements. Fleischmann and Pons proceeded to publish with an enormous discrepancy between their claimed excess heat and fusion products. This should have been a lightning rod for caution! Instead Fleischmann and Pons rushed into print announcing something with far reaching implications in nuclear physics that they didn't even understand. The premature publication on cold fusion by the two University of Utah chemists under such circumstances appears to others to have been motivated by a craving for fame, notoriety, dollars and patent rights.

Publication by Press Conference

An early press conference to announce a new scientific discovery may at times be justified under certain conditions, but it certainly is not the conventional method of communication in science since it circumvents the normal peer review process. If authors choose to communicate their results by press conference, they must be convinced that their results are correct and be able to back up their results with supporting evidence from all the obvious experimental checks. Pons and Fleischmann failed to carry out a number of even the more elementary tests and cross-checks. When questioned about their results with ordinary light water their answers were noninformative and subject to ambiguous interpretations. Completely inadequate nuclear-particle detectors and techniques were employed in their searches for fusion products. As a consequence, Fleischmann and Pons' neutron results were demonstrated to be artifacts.

Press conferences and press releases which present a new discovery in an unrealistic and overly optimistic tone serve to mislead the public. The University of Utah press conference left the incorrect impression that Fleischmann and Pons' experiments were very simple to perform and could easily be reproduced. Reporters responded to the University of Utah press conference by predicting that cold fusion would give an almost infinite supply of cheap and environmentally clean fuel leading to a revolution in the generation of energy. Some writers went so far as to envision a "fusion valley" in Utah that would become the economic mecca of scientific progress. However, publicity alone didn't make cold fusion true. What was missing was confirmation by independent researchers.

Since questions of carelessness and fraud in research are getting more and more public attention, it is especially critical that press conferences be accurate. Experiments must be clearly described and results honestly reported. In the case of the University of Utah press conference the initial facts reported about the discovery were incorrect. The claim made at the press conference that Fleischmann and Pons' device produced four watts of power for each watt consumed was inconsistent with that reported in their scientific article [*Journal of Electroanalytical Chemistry*, **261** 301 (1989)]. This article had been submitted[33] already on March 11, twelve days prior to

[33] Received by the journal on March 13, 1989.

the press conference, and it reported a power gain of a factor of ten less than the press conference! The popular press, however, continued to report the much larger, but erroneous power gain first publicized during the press conference. Initially, the very large power gain served to dramatize the discovery and assure the public that the effect was so large that it couldn't possibly be due to experimental error. In the long run, however, these highly exaggerated power claims served to discredit the University of Utah work when other scientists couldn't reproduce their results.

The timing of a press release or press conference must be such that sufficient data have been accumulated to minimize the risk of error. This usually requires that there is some degree of scientific consensus before a new discovery is widely reported in a press conference. Scientists who release their results directly to the public risk adverse reactions later if their results are shown to be erroneous.

Manipulation of the press by Fleischmann and Pons and University of Utah administrators contributed to the cold fusion furor. Exaggerated claims of excess heat and nuclear products were publicized without supporting information. As the facts became fully known, the initial claims were withdrawn or severely scaled back. Much of the circus atmosphere surrounding cold fusion could have been eliminated if the standard scientific procedure of peer review and publication in a reputable journal had been followed.

Publication of Primary Data

Researchers have the responsibility to publish their own experimental data. They should be in a position to explain and defend their results to other qualified experts. This is especially true when the reported data are controversial and directly contradict well-established scientific results in the literature. The description of experiments and results should be published in sufficient detail to give the expert reader the possibility of evaluating the significance of the claimed result. This standard scientific procedure was not followed by University of Utah scientists when reporting large amounts of ^{4}He in the gases evolving from Fleischmann and Pons' electrolytic cells, supposedly from D+D fusion.

On April 14, 1989, three weeks after the well-known University of Utah press conference, Walling and Simons, colleagues of Fleischmann and Pons, submitted a 'theoretical' paper to the *Journal of Physical Chemistry*. In this paper they reported, by way of a private communication from Pons and Hawkins, that the ^{4}He production in cold fusion cells was even larger than that required to account for all of the claimed excess heat. With this very preliminary and unsubstantiated evidence, Walling and Simons proceeded to construct their three-miracle 'non-theory' described in Chapter III. At this early stage in the history of cold fusion, ^{4}He was being promoted as the critical evidence for cold fusion. In the Editorial Comment section (a most unusual addition to a scientific paper) at the end of the Walling and Simons paper, it is recorded that one reviewer stated "it is of utmost importance to get the data presented in this paper into the public domain as quickly as possible." Although I support fast (and accurate) publication, the above advice to publish by private communication, data that directly contradicted well-established results in nuclear physics was not in the best long-term interest of science. Authors themselves have the responsibility to publish such findings along with supporting experimental evidence. It is completely unsatisfactory to introduce controversial and potentially important ex-perimental results into the literature through someone else's publication by way of a private communication.

On May 8, 1989 at the Los Angeles Electrochemical Society meeting, three weeks after the submission of the Walling and Simons paper, Fleisch-mann and Pons announced that their ^{4}He measurements were flawed. Nowhere in the Walling and Simons published paper (published on June 15, 1989) is this very serious mistake mentioned! The ^{4}He data, if true, would have been revolutionary. In actual fact, however, these data were completely erroneous, and no retraction has appeared. Presumably, since Pons and Hawkins did not publish their ^{4}He results, they felt no responsi-bility to retract the incorrect information put into print by Walling and Simons. On the other hand, Walling and Simons apparently felt that it was Pons' responsibility to set the record straight since they didn't add a note in proof warning the reader that the ^{4}He data were erroneous. This, of course, would have undermined the entire justification for publishing the Walling and Simons paper. The helium episode is an example of a worst case scenario of what can happen when data enter the scientific literature by way of private communication. When errors are discovered, they should be acknowledged immediately, preferably in the same journal.

Reproducibility in Science

The foundation of science requires that experimental results must be reproducible. Validation is an integral part of the scientific process. Scien-tists are obligated to write papers in such a way that the observations in them can be replicated. A set of instructions should be available to allow a competent and well-equipped scientist to perform the experiment and ob-tain essentially the same results. Replication in science is usually reserved for experiments of special importance or for experiments that conflict with an accepted body of work. The greater the implication of an experimental result, the more quickly it will be checked by other scientists.

In the case of cold fusion, confirmation proved elusive. After the March 23, 1989 press conference many scientists tried to duplicate Fleischmann and Pons' claims and concluded that whatever the University of Utah chemists had achieved in their electrolytical cells, it was not nuclear fusion. Even though a majority of scientists failed to replicate cold fusion, some scientists claimed partial confirmation, thereby fueling the cold fusion controversy. No one had been able to do a completely consistent cold fusion experiment with commensurate yields of excess heat and fusion products. Even so, fragmentary and sometimes contradictory evidence kept the cold fusion dream alive, at least in the eyes of the proponents.

As more and more groups at major universities and national laboratories were unable to replicate either the claimed excess heat or fusion products, proponents of cold fusion were quick to point out that the experiment was not done properly, one needed different size palladium cathodes, longer electrolysis times, higher currents, etc. Whenever the inability of qualified scientists to repeat an experiment is met by an onslaught of ad hoc excuses, beware. One of the important roles of a scientific paper is to provide directions for others to follow. Scientists establish priority for their discoveries by publishing a clear and well documented recipe of their experimental procedures. If a scientific paper does not include an adequate recipe allowing the skilled reader to reproduce the experiment, it is a warning that the authors' understanding of their experimental work is incomplete.

Cold fusion proponents introduced new dimensions into the subject of reproducibility in science. Some tried to turn the table on reproducibility to give irreproducibility some degree of respectability. This appraisal of cold fusion fooled no one. It remains that valid experimental results must be reproducible. A second aberration championed especially by Bockris was the assignment of a very different value to experiments attempting replication. Only experiments that reproduced some fragmentary bits of cold fusion were to be taken seriously since he declared that experiments obtaining negative results required no special skills or expertise. This philosophy led proponents of cold fusion when organizing conferences to invite mainly papers reporting positive results. This aberrant procedure is incompatible with the scientific process and was usually viewed negatively by the scientific community and the media.

Scientific Isolation in Research

Scientific research does not flourish in isolation. Instead, in order to succeed, the scientific process usually takes place within a broad social and historical arena where researchers exchange their ideas and observations. This interchange and interaction with colleagues is crucial in the modification of one's ideas and the sharpening and reformulation of a scientific advance. Failure to communicate suppresses the free and open debate by which scholars teach one another. Scientific breakthroughs usually result from the intense efforts of researchers working at the frontiers of their discipline, cognizant of all that has gone on before. Since isolation in science is always a handicap and never an advantage, it is difficult to understand some statements made by University of Utah President Peterson in his testimony before the House Committee on Science, Space and Technology (see Chapter V). For example, in his discussion of why the fundamental discovery of cold fusion was made by chemists in Utah rather than at an eastern "establishment" university, Peterson stated "There in fact may be something valuable in isolation from more traditional centers." The isolation argument was used also to suggest that chemists had an advantage when looking at a problem traditionally reserved to physicists. His view that science thrives in isolation makes good material for novels but is not a realistic view of the modern scientific enterprise.

The whole cold fusion saga is a striking illustration of what happens when science is done in isolation. Fleischmann and Pons' claim of having observed nuclear fusion at room temperature lacked credibility from the beginning because not only were their results contrary to all understanding of nuclear reactions, their reported neutron data were artifacts. There is no justification, be it potential patents, fame or riches, for doing science in this modern era on a major university campus in isolation from colleagues who would have had an important impact on the work. All evidence indicates that University of Utah administrators fostered the isolation of Fleischmann and Pons' cold fusion research from nuclear scientists both on and off campus. When after the press release, offers came to the university to provide equipment at no cost to assist in the verification process, Peterson's response was that Pons was in the process of setting up the trade of people and equipment with Los Alamos. For various reasons these collaborations with other groups did not happen and isolation prevailed.

A similar occurrence of a discredited discovery happened in 1972 at the University of Utah. Edward Eyring, Professor of Chemistry, reported the discovery of the x-ray laser, working in isolation from expert laser scientists. In a short time it was found that Eyring's claim was a mistake, and was dubbed the "Utah Effect" by scientists outside Utah. In referring to cold fusion, some wrote that the "Utah Effect" had struck again. The above examples serve to illustrate the grave dangers encountered when one attempts to do scientific research in isolation.

Control of Information

Open communication is particularly important in advancing scientific knowledge. Any attempt to control information can backfire. A scientist who refuses to divulge research data and materials to qualified colleagues runs the risk of not being trusted or respected. Many instances occurred during the short history of cold fusion where information was withheld or falsified.

In the early weeks after the University of Utah press conference and following the admission that the measurements of large amounts of helium in the exit gases from the electrolytic cells were flawed, there was a clamor for assessing the helium content in cathodes claiming to have produced excess heat. Such experimental measurements could in principle have been done in a few days by any number of laboratories with experienced groups in mass spectrometry. Fleischmann and Pons response to requests at Dallas was to tell those offering to do the measurements that such assays were already underway. Some weeks later at Los Angeles they replied that steps were being taken to have their cathodes analyzed. When no results were forthcoming, pressure from the scientific community led the two University of Utah chemists to orchestrate the so-called double-blind analysis of their cathodes for helium. Only one of the five palladium rods in this experiment was actually from an active cell claimed to have produced excess heat. During the exchange of the palladium rod histories and the data on the helium content (a crucial part of the double blind experiment), Pons reneged on the original agreement by holding back the power history of the single palladium rod which was claimed to have produced excess heat. Pons was now in an advantageous position to manipulate the results of the double-blind experiment knowing that his active palladium rod contained no helium above background. A month later he informed those involved in the double-blind experiment that his active cathode rod produced only milliwatts rather than the expected watts of power. Even so, the experimental helium data placed upper limits on any fusion having occurred as being many orders of magnitude below Pons' claimed milliwatt level of power for twenty-four days.

Fleischmann and Pons' reaction to the results of a University of Utah physics group that monitored their cells for fusion products over a five-week period was similar to their reaction on learning that their cells contained no helium. On learning that Salamon et al. found no fusion products, Pons again claimed that his cells weren't producing any excess power during this extended period (although he stated at an EPRI conference that his cells were producing low-level excess heat), except for a two-hour power excursion when the physicists' counters were dead. Through a judicious choice of experimental techniques, however, the University of Utah physicists were still able to set very low limits for any nuclear fusion occurring in Pons cells during the reported power excursion.

It was widely reported that the National Cold Fusion Institute (NCFI) was the recipient of a $500,000 gift from an anonymous donor. This supposedly generous gift was dangled before prospective donors as evidence of the credibility of cold fusion in the hope of raising substantial private and federal monies to supplement state funds. The source of the $500,000 was kept secret even from the director of the NCFI. In May, 1990 it was discovered by an outside probe that the funds had actually come from the University of Utah Research Foundation. President Peterson's misidentification of the source of the $500,000 caused a furor on the campus leading ultimately to the announcement that he would take early retirement at the end of the 1990–91 academic year. This is a case where rigid, secret control over the source of research funds was parlayed into a falsification of the source with the aim of catalyzing gifts from other donors. In the end, Peterson's scheme to keep the source of the funds secret backfired and led to a faculty vote of no confidence.

Secrecy in Basic Research

Basic research usually does not flourish in an environment of secrecy. This has been demonstrated once again in the cold fusion saga. On March 23, 1989 Fleischmann and Pons decided to go public with an announcement of what they considered to be a major scientific breakthrough in nuclear physics, an area in which they had no expertise. This was done without even seeking advice and opinions from the nuclear physicists on their own campus! Publicizing cold fusion to the world before discussing it with local nuclear physicists was a major mistake. It is particularly hard to understand such a decision in view of the fact that James Brophy, vice president for research and chief spokesman for the university on cold fusion, is a physicist by training. Before going public with a press conference announcing such a major claim that defied all previous evidence from nuclear physics, the University of Utah administrators should have insisted that the claim be reviewed by nuclear scientists both on and off campus. In order to make such a review meaningful, however, the university had to give other scientists access to the pertinent Fleischmann-Pons data. Rather than give out this information allowing others to evaluate the validity of the Fleischmann-Pons claim, the university administrators, fearing the loss of patent rights and royalties, chose to maintain secrecy right up to the time of the press conference. University of Utah President Peterson did call Hans Bethe, a leading nuclear theorist, and tell him about the forthcoming press release. By Peterson's own testimony, Bethe advised caution and a delay in the announcement; however, this expert advice was ignored.

In a modern university, it is most surprising how the Fleischmann-Pons research on nuclear fusion could have remained secret from their physics colleagues on campus for the five years that Fleischmann and Pons claimed to have worked on cold fusion. To have maintained this secrecy right up to the exciting stage of a press conference is beyond comprehension. After the press conference, physicists at the University of Utah were as puzzled by the claim as were their colleagues at other institutions who called them for information. Hence, they could be of no assistance in transmitting local recipes for how to induce nuclear fusion in a jar at room temperatures.

Discovery by Outsiders

In many respects this section parallels the earlier discussion where scientific isolation in research is considered. Fleischmann and Pons were electrochemists reporting a fundamental new discovery in nuclear physics, a discipline extensively researched during the last half century. On rare occasions self-taught outsiders make a discovery in an area largely unknown to them, but this is indeed rare. Most fundamental discoveries are made by persons intimately familiar with their research discipline because they not only know the subject matter of the field but also know the pitfalls and traps and have made many of the obvious mistakes.

As novices in nuclear physics, Fleischmann and Pons fell into several traps that might have been avoided had they consulted experts in nuclear physics. For example, it has been clearly shown that they lacked the equipment and expertise to assay their cells for the production of fusion products. Even assuming their reported neutron measurements were correct, their fusion product yield was many orders of magnitude less than their claimed excess heat. This colossal discrepancy immediately tipped off nuclear scientists that there was no evidence to assign the claimed excess heat to nuclear fusion. Fleischmann and Pons, however, were more impressed by their calorimetry, a specialty supposedly known to them, than they were by all previous nuclear physics data on D+D fusion. They reasoned that deuterons could be squeezed closely enough together inside palladium to have fusion occur. It didn't seem to worry them that they had to invoke an unknown nuclear process or processes to account for their claimed excess heat. In fact Fleischmann and Pons did not even seem to realize that they were novices in nuclear physics! In their initial press release (see Appendix I), Fleischmann stated, "We realize we are singularly fortunate in having the combination of knowledge that allowed us to accomplish a fusion reaction in this new way"; and Pons stated, "Without our particular backgrounds, you wouldn't think of the combination of circumstances required to get this to work."

Most scientists in the early months paid more attention to the much more modest claim about cold fusion made by physicist Jones and his collaborators. This stemmed from the fact that their publications and lectures displayed a keener awareness of the potential pitfalls and errors in detecting very low levels of fusion products. Although this claim has been treated as a genuine scientific claim, there are many skeptics questioning its validity based on uncertainties in the background.

Outsiders have the advantage in terms of the public's perception of their research accomplishments in an unfamiliar field. This was especially true in fusion research where the experts had spent billions of dollars with no payoff in sight for controlled fusion energy. When Fleischmann and Pons announced their claimed breakthrough in an area in which they were novices, it is no wonder that such news catalyzed the two opposite emotional outbursts witnessed at the Dallas ACS and Baltimore APS meetings. The natural tendency is to discount the skeptics and compulsive naysayers as disgruntled 'experts'. The University of Utah played this game skillfully labelling the skeptics as part of the 'eastern establishment' with conflicting interests in hot fusion. The University of Utah officials even debunked the warnings coming from their own physicists as simple academic jealousy.

Some outsiders to nuclear physics working on cold fusion went so far as to claim conventional nuclear physics was only applicable in gases. Hence, by fiat, nuclear fusion of deuterium in palladium lattices was assumed to follow a new type of nuclear physics allowing for greatly enhanced fusion rates, exotic branching ratios and concealed nuclear reaction products. Walling and Simons invoked all of these miracles to 'explain' Pons' claimed ^{4}He production, a claim retracted five weeks before the Walling-Simons paper was published. Walling and Simons, using their knowledge as chemists, made a preposterous extrapolation of the molecular level to the nuclear level in their paper. They concluded, for example, that the reaction energy from the reaction branch $D + D \rightarrow {}^4He +$ gamma went entirely into lattice heat rather than radiation (the 23.8–MeV gamma ray). For the high-energy nuclear case, there cannot be any appreciable coupling between the photon and the atomic electrons, and internal conversion or any related process cannot take place at anywhere near the rate postulated. This is a striking example of how far science may be set back by outsiders when indiscriminately applying their models to another discipline. The erroneous Walling-Simons paper served only to fuel the cold fusion controversy!

Lobbying Before Congressional Committees

Approximately one month after announcing cold fusion to the world, the University of Utah participated in a hearing before a Congressional Committee promoting their claimed discovery to Congress. The University of Utah delegation included the principals Fleischmann, Pons and President Peterson and, in addition, Ira C. Magaziner and Gerald S.J. Cassidy, the heads of high-powered lobbying firms. Their purpose in Washington was obvious; to lobby for an immediate $25 million (and an eventual $125 million) to supplement $5 million of promised Utah state funds to start a National Cold Fusion Institute (NCFI). Many scientists were appalled that the University of Utah delegation would appear before the House of Representatives' Committee on Science, Space and Technology (see Chapter V) exuding excitement about cold fusion and requesting funds before their cold fusion claim had even been confirmed by other independent scientists.

It has now become rather common practice for colleges and universities to go directly to Congress seeking funds, thereby avoiding competitive grant proposals and the peer review process. This end-run for funds is defended by second-tier universities as a mechanism for them to obtain a more equitable share of scarce federal research funds. On the other hand, top-rank universities argue that federal funding should be allocated on the basis of quality and these funds should be used to support the very best research. The tendency for more and more universities to hire lobbyists to persuade members of Congress to allocate federal funds directly for home-state projects is alarming and needs to be addressed. Firms such as Cassidy and Associates, while collecting large fees, have played a major role in assisting academic institutions to obtain a share of these 'pork-barrel' funds. This practice wastes limited research funds and serves to weaken the research infrastructure in the United States, thereby impairing our ability to compete internationally.

In view of the prevalent practice of U.S. universities hiring lobbyists and the success they have enjoyed, it was not surprising that Cassidy and Associates accompanied the University of Utah team around Congress in its hard sell of an unproven discovery. This is a blatant example where lobbying for research funds has been shown to be an insidious practice. The University of Utah delegation was putting pressure on members of Congress to release funds rapidly for a research project that already at that time could not be reproduced by a number of expert research groups. Fortunately, Congress in its wisdom refused to allocate funds for the proposed National Cold Fusion Institute. The University of Utah lobbying episode raises serious questions about the way research funds are obtained through lobbyists without any check on the merit of the science by expert referees. Congress

229

itself has a role in stopping the practice of earmarking academic-science funds for non-peer-reviewed projects at favorite universities. 'Pork barrel' science is a bad policy in good times but an outrage in times of short budgets. The cold fusion fiasco is a stern reminder about the possible dangers of a system where research funds are obtained by going directly to Congress.

Funding Large Initiatives

Utah State's legislature moved quickly after Fleischmann and Pons' press conference to establish a $5 million fund to support and promote cold fusion research in Utah. This rapid action of the state legislature was taken with the stipulation that these funds would be released only after cold fusion was scientifically confirmed. To manage these funds, a Fusion/Energy Advisory Council was formed with members to be appointed by the governor. It was the responsibility of this council to determine what constituted scientific confirmation. Although there is little information available about the council's deliberations on this key question, they eventually sought advice from two outside witnesses, John Bockris of Texas A&M University and Robert Huggins of Stanford University. Most would agree that they are the two staunchest supporters of the Fleischmann-Pons phenomenon in the United States.

On July 21, 1989 the Advisory Council voted unanimously to accept the Fleischmann-Pons claims on cold fusion as confirmed. The Council's deliberations were taking place concurrently with our DOE/ERAB panel's work on our interim report, completed at the July 11–12, 1989 meeting. Our panel found at that time no convincing evidence that useful sources of energy would result from the Fleischmann-Pons phenomenon. In fact the panel's interim report went on to say that the evidence for the discovery of a new nuclear process termed cold fusion was not persuasive. To my knowledge the Utah Advisory Council made neither an effort to learn about our panel's work and it's negative evaluation of cold fusion nor did they consult anyone from the large number of research groups getting negative results.

The type of peer-review conducted by Utah's Advisory Council is inappropriate when making decisions on research funding, especially so when large sums of money are involved. Scarce research funds should be used to support the very best research. When this is not done, everyone loses. Having spent the $5 million of state funds, no confirmatory evidence for cold fusion has resulted. This is an expensive lesson to be learned for those responsible for the way research is funded.

Patents and Revenues from Basic Science

It has been stated that the two University of Utah chemists announced their discovery of cold fusion before they were ready to do so. It is certainly true that the announcement was made before they fully understood their results, requiring them to state that the bulk of their reported heat was due to an unknown nuclear process or processes. In fact the disparity between their initially reported heat and fusion products was approximately a factor of a billion! Under these circumstances, what caused Fleischmann and Pons to go public so soon? Various evidence suggests that they did so at the request of University of Utah officials.

The scenario that is emerging suggests that the University of Utah administrators became concerned that the news of Fleischmann and Pons' research was about to leak out. Worried that this leak would jeopardize the University of Utah's case for commercializing cold fusion, officials requested that the two chemists prepare a preliminary paper on their research and to immediately hold a press conference even before the paper was published. This accelerated time schedule resulted in a paper that was riddled with errors and eventually requiring the publication of an unusual amount of errata including a third author.

Administrators at other U.S. universities agreed that the way the University of Utah officials handled the cold fusion phenomenon was an aberration and not the way that most universities deal with scientific discoveries and the results of basic research. Dr. J.L. Cohon, Vice-Provost for research at Johns Hopkins University said "It was wrong the way it was done . . . [I]t calls into question the credibility of all research organizations and research universities" (*The Chronicle of Higher Education*, May 17, 1989). A more normal procedure for protecting the commercialization of new technologies is to delay publication of research papers allowing time for patents to be filed. Then, after the research is published a public announcement might be made. At the University of Utah this normal sequence of events was reversed.

One of the serious problems with both the University of Utah's press conference and preliminary paper was that they lacked many of the important details necessary to allow others to either evaluate or duplicate the cold fusion phenomenon. Many scientists, research administrators and a host of others were angered and frustrated by what appeared to them an undue concern for commercialization by the University of Utah. Most would have preferred that the debate on cold fusion take place in the scientific literature rather than in the public press. Instead the University of Utah chose to distort the normal scientific procedure by limiting access to their data in order to protect patent rights. Time and time again the University of Utah

231

representatives justified their actions on the basis of their lawyers' concern for protection of the institution's intellectual property rights. This should not have been a concern, as stated by the University of Houston counsel J. Scott Chafin, who was involved in the commercial aspects of major superconductivity discoveries there. He stated, "I don't find any fault in trying to protect intellectual property . . . [W]hat I fault is selectively revealing what the science is. One can disclose all the science and still protect the intellectual property" (*ibid.*).

The University of Utah's handling of the cold fusion episode appeared from the outside to be unduly driven by the grandiose vision of large amounts of royalties. This caused an undue emphasis on secrecy, intellectual property rights and patents at a time when the primary emphasis should have been on verification. Only when the science is validated, should one proceed to the stage of applying for patents. The University of Utah's violation of this usual scientific protocol has damaged their reputation and wasted large amounts of scarce resources, and should serve as a lesson for all research universities.

The Press and Basic Science

Scientists and journalists often have different points of view about what is newsworthy.[34] Scientists usually publish in peer-reviewed journals and strive to fit their discoveries into the larger framework of information in their special discipline, emphasizing the continuity and cumulative nature of science. Interesting contributions to science are usually those that advance the field as a whole in steady small steps, although on rare occasions major advances occur like the recent discovery of high-temperature superconductivity. On the other hand journalists tend to focus on striking single events that are often dramatic or even controversial, such as cold fusion. Journalists also tend to emphasize personalities and personality conflicts.

When following the normal procedure of scientific publication, the information is not available for release to the public until after the manuscript has been prepared, submitted for publication, passed through the peer review process and released by the journal in printed form. At this point the journalist's task is to select and simplify the complex material in such a way as to make it interesting to the reader while retaining its scientific integrity. This is a very difficult task because scientists and journalists have different styles of communication and different norms of objectivity. Furthermore, scientists and journalists sometimes even differ on the appropriate role of the press. In the case of cold fusion the above procedure was markedly

[34] See, for example, the article by Dorothy Nelkin, *Physics Today*, November (1990).

altered with the initial information released during a press conference before any published paper was available for backup data. Hence, the journalists, along with everyone else, were at the mercy of the University of Utah press office for their information.

The University of Utah administrators, as well as Fleischmann and Pons, used the press in the early weeks as a conduit to disseminate their selected choice of information to the public. For example, it was from their initial press conference on March 23, 1989 that came the oft-heard phrase of "four watts output for one watt input."[35] Other directed messages in the early releases were the simplicity of the experiments and the many times that Fleischmann and Pons had repeated the experiments over the last five years, sometimes for stretches of 100 hours. All of this information served to give the public a highly positive and optimistic view about cold fusion. Since most people learn about science from what they read in the press, the majority of the early stories gave the public the highly biased view that a breakthrough had occurred in scientists' quest for fusion energy.

As more and more journalists began probing the University of Utah claims, the neatly packaged 'breakthrough' announced at the initial press conference began to unravel. Fleischmann and Pons gradually withdrew from the the public eye refusing interviews and avoiding the media. They also refused to be interviewed during any of the official press conferences at the First Annual Conference on Cold Fusion. After the scientific sessions they would leave quickly to escape interaction with the many reporters covering the conference. At this time they went so far as to publish a criticism of the press in the *Deseret News* entitled, "Press Should Separate Facts, Opinion." This article is primarily a caustic criticism of *Nature* and its editors but the piece has a far larger implication as it states, "We have used Nature simply by way of illustration – we have plenty of vitriol in store for other journals and newspapers." When much of the press no longer served as a conduit to popularize the University of Utah's perspective on cold fusion, animosity developed between members of the two organizations.

When speaking about the press one must remember that it stands for a highly complex organization consisting of science writers, editors, etc. who approach their task in a multitude of ways. Although some science writers tend only to elucidate and explain complex phenomena in a way understandable to the public, others go beyond this and challenge their sources of information. When the latter is done, the public is better served, given the

---

[35] This claimed large power gain was completely inconsistent with the paper Fleischmann and Pons had submitted to the *Journal of Electroanalytical Chemistry* several days previously. In their paper they claimed approximately 1.3 watts output power for 1 watt input.

importance of science in society. There were many examples of excellent investigative journalism during the cold fusion saga, where reporters made an effort to present a balanced view of cold fusion. On the other hand some science writers chose to stay on the "cold fusion team" and report principally positive claims without checking with the many scientists getting negative results.

The Scientific Process

The whole cold fusion fiasco serves to illustrate how the scientific process works. However, seldom do far-out claims receive the amount of national and international attention given to cold fusion. Scientists are real people and errors and mistakes do occur in science. These are usually detected either in early discussions of ones research with colleagues or in the peer-review process. If mistakes escape notice prior to publication, the published work will come under close scrutiny by other scientists, especially if it disagrees with an established body of data. The greater the implication of a result, the sooner it will be reexamined. Scientific results, if valid, must be reproducible. When errors are discovered, acknowledged and corrected, the scientific process moves quickly back on track, usually without either notice or comment in the public press.

The scientific process is self corrective. This unique attribute sets science apart from most other activities. The scientific process may on some occasions move slowly, sometimes even along a circuitous path. The significant characteristic of the scientific method, however, is that in the end it can be relied upon to sort out the valid experimental results from background noise and error. This has been so firmly demonstrated again in the present case by showing that there is no evidence to support the claim of measurable amounts of heat energy coming from room temperature nuclear fusion. True progress must withstand the test of time.

When the news of cold fusion broke, the scientific establishment moved quickly to investigate the validity of the far-out claims which many thought were nonsense. The scientific establishment was not too arrogant to examine room temperature fusion when it came along, even though it contradicted well-established experimental and theoretical results in nuclear physics. There are occasionally surprises in science and one must be prepared for them. Within the first few weeks after the University of Utah press conference, several multidisciplinary research teams could not replicate any of Fleischmann and Pons' reported claims. Such teams were necessary to investigate quickly the cold fusion claims. Had Fleischmann and Pons formed an interdisciplinary team in their early experiments, cold fusion might have had a very short lifetime.

On the first week of May, cold fusion was the cover story of *Time*, *Newsweek* and *Business Week*, an unprecedented occurrence for a science story. Why did cold fusion generate so much national and international excitement? Following the many media stories about the erosion of our environment by the green house effect, acid rain, chemical and radioactive wastes, and the Valdez and Chernobyl disasters, the announcement on March 23, 1989 of the sudden possibility of an abundant, cheap and pollution-free energy source captured everyone's imagination. It was a dream come true, giving us new confidence of our technological competitiveness. Few were concerned or even aware that the University of Utah's publication-by-press-conference released the cold fusion story directly to the public, bypassing all the normal checks and controls of the scientific process. There was no manuscript available for evaluation, there had been no peer review of the science, in fact, the University administrators had not even consulted the nuclear physicists on their own campus. There was only an announcement that watts of excess energy had been produced by a nuclear fusion reaction at room temperature in a small electrolytic cell. The press conference did mention the observation of fusion products to reinforce the claim of nuclear fusion. There was no mention, however, that the fusion-product yield was more than eight orders of magnitude less than the claimed excess heat. Later it was shown by others that Fleischmann and Pons made their claim of nuclear fusion before having any solid evidence of having observed any fusion products.

The University of Utah's handling of cold fusion is a striking illustration of what happens when scientists circumvent the normal peer review process, when scientists use the press as a conduit to disseminate information about a claimed discovery in an unrealistic and overly optimistic tone, when scientists require too many miracles to account for their results, when research is done in isolation by scientists who are outside their field of expertise, when data are published by private communication rather than by those responsible, when administrators use potential royalties to force premature publication and when university administrators lobby for large federal funds before the science is confirmed. Cold fusion is an example of bad science where the normal rules and procedures of the scientific process were violated. One can only be amazed by the number of scientists who reported confirmation of cold fusion by press conference, only to follow later with a retraction or at least a confession of irreproducibility. Reproducibility is the essence of science. It has taken upwards of some fifty to one hundred million dollars of research time and resources to show that there is no convincing evidence for room temperature fusion. Much of this effort would not have been necessary had normal scientific procedures been followed. The idea of producing energy from room temperature fusion is destined to join N rays and polywater as another example of a scientific aberration.

The purpose for exposing the cold fusion episode is to show that serious mistakes do occur in science. It is important that we learn from these mistakes. I hope examples discussed in this book will give others new insights into the way science should be done. The general scientific enterprise is vibrant and healthy and has weathered the cold fusion flurry with only minor bruises and scratches. The cold fusion fiasco illustrates once again, as N rays and polywater did earlier, that the scientific process works by exposing and correcting its own errors.

APPENDIX I

University of Utah Press Release*

'SIMPLE EXPERIMENT' RESULTS IN SUSTAINED N-FUSION
AT ROOM TEMPERATURE FOR FIRST TIME

Breakthrough process has potential to
provide inexhaustible source of energy

SALT LAKE CITY – Two scientists have successfully created a sustained nuclear fusion reaction at room temperature in a chemistry laboratory at the University of Utah. The breakthrough means the world may someday rely on fusion for a clean virtually inexhaustible source of energy.

Collaborators in the discovery are Dr. Martin Fleischmann, professor of electrochemistry at the University of Southampton, England, and Dr. B. Stanley Pons, professor of chemistry and chairman of the Department of Chemistry at the University of Utah.

"What we have done is to open the door of a new research area," says Fleischmann. "Our indications are that the discovery will be relatively easy to make into a usable technology for generating heat and power, but continued work is needed, first, to further understand the science and secondly, to determine its value to energy economics."

Nuclear fusion offers the promise of providing humanity with a nearly unlimited supply of energy. It is more desirable than the nuclear fission process used today in nuclear power plants. Fusion creates a minimum of radioactive waste, gives off much more energy and has a virtually unlimited fuel source in the earth's oceans.

Nuclear fusion is also superior to traditional energy sources, such as coal, gas and oil, which can pollute the environment and eventually will be depleted. Using fusion for energy would reduce or even eliminate major causes of acid rain, the greenhouse effect and U.S. dependence on foreign oil.

Their findings will appear in the scientific literature in May.

Scientists worldwide have searched for more than three decades for the ability to create and sustain nuclear fusion reactions, which are thought to be the ideal energy source. In nature, the energy of stars, such as the sun, is supplied by nuclear fusion. All fossil fuels presently used on earth are simply storehouses of stellar nuclear fusion energy. Prior to the breakthrough research

* This document was embargoed for release on Thursday, March 23, 1989 at 1:00 p.m. Mountain Standard Time. I thank Professor Bruce V. Lewenstein for supplying this material from the Cold Fusion Archive Project at Cornell University.

237

at the University of Utah, imitating nature's fusion reactions in a laboratory has been extremely difficult and expensive.

In the Utah research, the electrochemists have created a surprisingly simple experiment that is equivalent to one in a freshman-level, college chemistry course. Conventional nuclear fusion research requires temperatures of millions of degrees, like those found in the sun's interior, to create a reaction. The Utah research, however, creates the reaction at room temperature.

In the experiment, electrochemical techniques are used to fuse some of the components of heavy water, which contains deuterium and occurs naturally in sea water.

Sea water provides essentially an unlimited source of deuterium. Even though it is present at only one part in 38,000, one cubic foot of sea water contains enough deuterium to produce 250,000 BTU of energy, which is equivalent to the energy produced from 10 tons of coal.

The scientists know their experimental result is fusion in an electrode because the generation of excess heat is proportional to the volume of the electrode. "This generation of heat continues over long periods, and is so large that it can only be attributed to a nuclear process," Fleischmann says. "Furthermore, side reactions lead to the generation of neutrons and tritium which are expected by-products of nuclear fusion." The device the researchers have constructed produces an energy output higher than the energy input.

Pons calls the experiment extremely simple. "Observations of the phenomenon required patient and detailed examination of very small effects. Once characterized and understood, it was a simple matter to scale the effects up to the levels we have attained."

The researchers' expertise in electrochemistry, physics and chemistry led them to make the discovery. "Without our particular backgrounds, you wouldn't think of the combination of circumstances required to get this to work," says Pons.

Some may call the discovery serendipity, but Fleischmann says it was more accident built on foreknowledge. "We realize we are singularly fortunate in having the combination of knowledge that allowed us to accomplish a fusion reaction in this new way."

The idea to attempt the innovative experiment was seeded in the late 1960s when Fleischmann conducted research on the separation of hydrogen and deuterium isotopes. The results were odd. His interpretation of the data indicated it would be worth looking for nuclear fusion reactions.

Later, in separate research, Pons looked at isotopic separation in electrodes and was puzzled at certain results. The two pondered the data and later discussed the findings on two memorable occasions, once when they drove together through Texas and later when they took a hike up Millcreek Canyon on the outskirts of Salt Lake City.

"Stan and I talk often of doing impossible experiments. We each have a good track record of getting them to work," says Fleischmann. "The stakes were so high with this one, we decided we had to try it."

The research strategy was concocted in the Pons' family kitchen. The nature of the experiment was so simple, says Pons, that at first it was done for the fun of it and to satisfy scientific curiosity. "It had a one in a billion chance of working although it made perfectly good sense."

The two performed the experiment and had immediate indication that it worked. They decided to self-fund the early research rather than try to raise funds outside the University because, says Pons, "We thought we wouldn't be able to raise any money since the experiment was so farfetched."

Working late into the night and on weekends at Pons' University of Utah laboratory, the two improved and tested the procedure throughout a five-and-half year period.

"We hope we'll be able to work with others to develop this into a useable technology for generating heat and power for the world," says Fleischmann. "The process is clean and indications are it will be economical compared to conventional nuclear systems."

Fleischmann has written more than 240 articles in the electrochemical, physics, chemistry and electrochemical engineering fields during his 40-year career, and is regarded as one of the leading electrochemists in the world. He is a fellow of the Royal Society of England. He was awarded a medal for Electrochemistry and Thermodynamics by the Royal Society of Chemistry in 1979; the Olin-Palladium Medal of the Electrochemical Society in 1985; and the Bruno Breyer award by the Royal Australian Chemical Society in 1988. He earned a doctorate in chemistry at London University in 1951.

He and Pons have collaborated on 32 articles.

Pons has authored more than 140 articles and lectured throughout the United States, Canada and Europe. He earned a bachelor of science degree at Wake Forest University, Winston-Salem, North Carolina in 1965 and a doctorate at the University of Southampton, England, in 1979. He is originally from Valdese, North Carolina.

Working on the project with the two scientists is University of Utah graduate student Marvin Hawkins from LaJara, Colorado.

The fusion technology is owned by the University of Utah which has filed patent applications covering the technology. Information about commercial aspects of the technology development can be obtained from Dr. Norman Brown, director of the University of Utah Office of Technology Transfer. 801–581-7792.

The researchers are grateful for the encouragement of the United states Office of Naval Research, their respective universities, families and colleagues.

APPENDIX II

Energy Research Advisory Board Cold Fusion Panel

*John R. Huizenga, Co-Chairman
Tracy H. Harris Professor of
  Chemistry and Physics
University of Rochester

*Norman Ramsey, Co-Chairman
Higgins Professor of Physics
Harvard University

Allen J. Bard
Norman Hackerman-Welch Regents
Professor of Chemistry
University of Texas

Jacob Bigeleisen
Distinguished Professor of Chemistry
SUNY, Stony Brook

Howard K. Birnbaum
Professor of Materials Science
Materials Research Laboratory
University of Illinois

Michel Boudart
Professor of Chemical Engineering
Stanford University

Clayton F. Callis
President
American Chemical Society

*Mildred Dresselhaus
Institute Professor
MIT

Larry R. Faulkner
Head, and Professor,
Department of Chemistry
University of Illinois

T. Kenneth Fowler
Professor of Nuclear Engineering
University of California, Berkeley

Richard L. Garwin
IBM Fellow & Science Advisor
  to the Director of Research
IBM Corporation

*Joseph Gavin, Jr.
Senior Management Consultant
Grumman Corporation

William Happer, Jr.
Professor of Physics
Princeton University

Darleane C. Hoffman
Professor of Chemistry
University of California, Berkeley

Steven E. Koonin
Professor of Theoretical Physics
CALTECH

Peter Lipman
US Geological Survey
Denver, CO

Barry Miller
Supervisor, Analytical Chemical
Research Department
AT&T Bell Laboratories

David Nelson
Professor of Physics
Harvard University

*ERAB Members

John P. Schiffer
Associate Division Director, Physics
Argonne National Laboratory, and
Professor of Physics
University of Chicago

*John Schoettler, ex officio †
Independent Petroleum Geologist

*Dale Stein
President
Michigan Technological University

Mark Wrighton
Head, and Professor,
Department of Chemistry
MIT, Building 6335

STAFF

Thomas G. Finn
Executive Director, ERAB

David Goodwin
Panel Technical Advisor

William Woodard
Secretary, Cold Fusion Panel

† In July, 1989, John Landis, Senior Vice President of Stone & Webster Engineering Corporation became the Chairman of ERAB replacing John Schoettler as an ex officio member.

Appendix III

Chronology of the Cold Fusion Saga

August, 1988	Fleischmann and Pons prepared a research proposal on electrolytic fusion. In late August they submitted it to Dr. Ryszard Gajewski, project director of the Advanced Energy Projects Division of the Department of Energy.
September, 1988	Gajewski selects Jones as one of five referees for the Fleischmann and Pons proposal. Jones submitted his report in late September. An intense competition developed between Fleischmann and Pons and Jones *et al.*
February 3, 1989	Deadline for receipt of Jones' abstract for the May 1–4, 1989 Baltimore APS Meeting
February 23, 1989	Fleischmann and Pons visited Jones at BYU and toured his laboratory. Jones showed his neutron data to Fleischmann and Pons and indicated that the BYU group felt they had enough information to merit journal publication.
March 6, 1989	Presidents of the University of Utah and BYU along with Fleischmann, Pons, Jones and others met at BYU. It was reported that an agreement was reached to mail simultaneously manuscripts on cold fusion from each institution to *Nature* on March 24 from Salt Lake City.
March 11, 1989	Fleischmann and Pons submitted a paper on their cold fusion research to the *Journal of Electroanalytical Chemistry* (unknown to Jones). Received at Journal office on March 13; in revised form on March 22.
March 13, 1989	First United States patent application by the University of Utah; other filing dates are March 21, April 10, April 14, April 18, May 2 and May 16, 1989.
March 13, 1989	Fleischmann informs David Williams of the Harwell Laboratory (UK) of their results and Williams embarks on a large scale multidisciplinary program on cold fusion.

March 21, 1989	University of Utah made critical decision to go public with a press conference without informing anyone at BYU.
March 22, 1989	*Financial Times* in London breaks the cold fusion story with Fleischmann's assistance.
March 23, 1989	Fleischmann and Pons hold a press conference in Salt Lake City announcing the discovery of room temperature nuclear fusion. Claimed detection of excess power (4 watts output for 1 watt input), neutrons, tritium and gamma rays.
March 24, 1989	On request Brigham Young University handed out a statement summarizing the observations of very low intensity neutrons in room temperature fusion by Jones *et al.* Their manuscript had been faxed to *Nature* so that it would have a March 24 submission date.
March 24, 1989	Teller phoned Pons for supplementary information. Pons sent him a preprint of their manuscript. Teller comments favorably to the press.
March 28, 1989	Fleischmann presents a colloquium at Harwell and discusses the University of Utah experiments with Williams and Harwell scientists.
March 31, 1989	Pons gives a seminar at University of Utah. Jones gives a physics department colloquium at Columbia. Fleischmann lectured at CERN.
April 10, 1989	Publication of the first paper by Fleischmann and Pons [*J. Electroanal. Chem.* **261** 301 (1989)] claiming excess heat, neutrons, tritium and gamma rays from room temperature fusion of D+D in small electrolytic cells with palladium electrodes. Marvin Hawkins name was omitted from the above paper and later added as part of the extensive errata!
April 12, 1989	Special session on cold fusion featuring Pons at the Dallas American Chemical Society Meeting attended by 7,000 chemists in the Convention Center. Sometimes called the "Woodstock of Chemistry".
April 12, 1989	Jones and Fleischmann speak at a world conference to discuss cold fusion organized by Ettore Majorama Centre for Scientific Culture in Erice, Sicily.

April 13, 1989	Seaborg summoned to Washington to meet with Watkins, Sununu and President Bush; Ten national laboratories instructed to do cold fusion research and an ERAB panel to be formed to advise the Secretary of Energy.
April 14, 1989	Walling and Simons submit theoretical paper to *J. of Phys. Chem.* claiming, by way of a private communication from Pons, that ^{4}He production is commensurate with excess heat.
April 18, 1989	Professor E. Amaldi spoke at the National Academy of Science meeting about Frascati results on non-equilibrium or dynamic origin of cold fusion.
April 26, 1989	Hearing on cold fusion before the House Committee on Science, Space and Technology in Washington. The University of Utah delegation included Fleischmann, Pons, President Peterson and the heads of two lobbying firms, Ira C. Magaziner and Gerald S.J. Cassidy. The University of Utah was requesting an immediate $25 million (and an eventual $125 million) to supplement a promised $5 million of state funds for a National Cold Fusion Institute before the science was confirmed.
April 27, 1989	Publication of the first paper by Jones et al. [*Nature* **338** 737 (1989)] claiming very low-level neutron emission in cold fusion.
May 1 and 2, 1989	Special sessions on cold fusion at the Baltimore American Physical Society meeting. Several groups presented negative results. As a result of this meeting, the public's perception of cold fusion was changing and becoming more skeptical.
May 8, 1989	Special session on cold fusion at the Los Angeles meeting of the American Electrochemical Society. No new data from the University of Utah. Fleischmann announced that their results both on ^{4}He in their cells effluent gases and on neutrons were flawed.
May 18, 1989	Publication of the paper by Petrasso *et al.* [*Nature* **339** 183 (1989)] showing that the Fleischmann and Pons neutron results were artifacts. Very damaging to the credibility of Fleischmann and Pons' claim of cold fusion.
May 23–25, 1989	Workshop on Cold Fusion phenomena in Santa Fe hosted by the Los Alamos National Laboratory

and attended by 400 participants (Fleischmann and Pons boycotted this meeting). Attention given to neutron bursts reported by Menlove and Bockris' tritium reported by Wolf. Many negative results reported. Impressive were the closed-cell calorimetry results from Canada and the low limits placed on cold fusion neutrons by Gai and DeClais. DOE/ERAB Panel holds its first meeting. Jones and Gai agree to do a joint experiment at Yale.

June 2, 1989 DOE/ERAB panel members visit the laboratories of Fleischmann and Pons and Wadsworth. Panel failed to get cell calibration data from Pons.

June 13, 1989 DOE/ERAB panel members visit BYU. Some concern expressed about whether the adopted neutron background had been corrected for changes in barometric pressure.

June 15, 1989 Harwell called a press conference to announce that it was terminating its research on cold fusion after its large multidisciplinary team of scientists found neither excess heat nor fusion products.

June 19, 1989 DOE/ERAB panel members visited Texas A&M University groups of Appleby, Bockris, Martin and Wolf. Panel focussed on the tritium results.

June 20, 1989 DOE/ERAB panel members visited California Institute of Technology where a group of 15 chemists and physicists, directed by Lewis and Barnes, found neither excess heat nor fusion products.

June 22, 1989 DOE/ERAB panel members met in Washington, D.C. to discuss all available information on cold fusion and to formulate a preliminary outline of its interim report.

July 6, 1989 DOE/ERAB panel members visit Huggins group at Stanford and McKubre group (Stanford Research Institute) at EPRI.

July 6, 1989 Publication of the paper by Gai et al. [Nature 340 29 (1989)] pushing the upper limit on the neutron intensity from cold fusion more than an order of magnitude below Jones et al.

July 11 & 12, 1989 DOE/ERAB panel members met in Washington to complete its interim report. Our panel concluded "that the experiments reported to date do not present convincing evidence that useful sources of energy will result from the phenomena attributed to cold fusion. Indeed, evidence for the discovery of a new nuclear process termed cold fusion is not persuasive. Hence no special programs to establish cold fusion research centers . . . are justified . . ."

August 7, 1989

National Cold Fusion Institute (NCFI) in Salt Lake City officially opened with state funds.

August 17, 1989

Publication of the paper by Lewis *et al.* [*Nature* **340** 525 (1989)] reporting neither excess heat nor fusion products from cold fusion.

September 15, 1989

Deadline set by DOE/ERAB panel for receipt of cold fusion reports from national laboratories, universities, and industries in the United States and abroad. An extremely large amount of material was received.

October 13, 1989

DOE/ERAB panel members met at O'Hare Airport in Chicago to review reports of sub-panels and to formulate the structure and content of our final report.

October 16–18, 1989

NSF/EPRI meeting on cold fusion by invitation only. Most of the 50 attendees were believers. Sensational (but erroneous) results reported on isotopic enrichment on the surface of the cathode, leading Teller to postulate a new particle which he named the meshugatron. Organizers gave positive press release in anticipation of our DOE/ERAB panel's negative final report.

October 30 and 31, 1989

DOE/ERAB panel members met in Washington, D.C. to complete its 69-page final report. The fifth conclusion captures the negative tone of our report "Nuclear fusion at room temperature, of the type discussed in this report, would be contrary to all understanding gained of nuclear reactions in the last half century . . ."

November 3, 1989

Interim Director Rossi of the NCFI issued a statement to the press very similar to the conclusions reached by our DOE/ERAB panel. University of Utah then restricted NCFI scientists interaction with the press and shortly thereafter Rossi resigned.

November 23, 1989

Publication of the paper by Williams *et al.* [*Nature* **342** 375 (1989)] reporting neither excess heat nor fusion products from cold fusion. Our DOE/ERAB panel had a preprint of this paper well before our final report.

December, 1989
Publication by Bhabha Atomic Research Center of report BARC–1500 giving positive results, especially for tritium. Our DOE/ERAB panel had an early version of this report during summer of 1989.

February 1, 1990
Fritz C. Will takes over as the first permanent director of the NCFI. Verification of Fleischmann-Pons phenomenon a priority.

March 12, 1990
University of Utah's international patent application was filed and is based on seven applications filed in the United States between March 13 and May 16, 1989. One surprising feature is that the list of inventors has, in addition to Fleischmann and Pons, Walling and Simons.

March 29, 1990
Publication of the paper by Salamon et al. [Nature **344** 401 (1990)]. This University of Utah physics group monitored Fleischmann and Pons electrolytic cells between May 9 and June 6, 1989 and found no evidence for any fusion products. In April, 1990 the authors each received a letter from Pons' lawyer C. Gary Triggs threatening them with legal action if their paper was not retracted.

March 29–31, 1990
The First Annual Conference on Cold Fusion in Salt Lake city. Most of the over 200 participants were believers and all the talks were positive. It was an United States production with a token number of foreign papers and participants. Large representation of the media, who were mostly critical including the local reporters. Organizers tried to control press conferences by including only those scientists who were proponents of cold fusion. Fleischmann told me that he thought the unknown nuclear process was the fission of palladium. It was a meeting like no other scientific meeting that I've ever attended.

May, 1990
Taubes and Fitzpatrick discovered that University of Utah President Peterson had falsified the origin of a $500,000 gift to the NCFI.

June 1, 1990
President Peterson admitted that his handling of the above large gift was a mistake.

June 4, 1990
Resolution passed by the Academic Senate questioning President Peterson's leadership.

June 11, 1990	President Peterson announced he would retire at the end of the 1990–91 academic year to end the unproductive controversy on the University of Utah campus.
June 15, 1990	Article by Gary Taubes [*Science* **248** 1299 (1990)] discussing Texas A&M University's handling of Bockris' tritium results. Raises the question about possible fraud or sloppy science.
October 22–24, 1990	Meeting on Anomalous Nuclear Effects in Deuterium/Solid Systems held at BYU. Little new by way of positive claims reported.
October 23, 1990	Pons missing, his home was up for sale and his telephone disconnected. On October 24 his attorney sent a one-sentence fax message requesting a sabbatical.
November 7, 1990	Scientific review of NCFI by an external committee of four members outside Utah. They concluded "that neither the institute's various experiments nor those anywhere else have firmly established the existence of cold fusion". This conclusion is the same as that reached by our DOE/ERAB panel some one and a half years earlier.
January 9, 1991	The University of Utah announced the resignation of Pons from his tenured faculty position effective January 1. He was appointed as a research professor for a limited period of 18 months.
February 1, 1991	The Utah Advisory Council gave Pons this date as a deadline for turning over all of his research notebooks to Council member Hansen.
March 11, 1991	Fritz C. Will resigns from the NCFI's board of directors claiming that Fleischmann and Pons did not turn over all their experimental data as required under a written agreement on February 1, 1991.
June 29–July 4, 1991	Second Annual Conference on Cold Fusion was held at Villa Coma, Italy. International Advisory Committee was made up of well-known strong believers in cold fusion.
June 30, 1991	National Cold Fusion Institute closes.

INDEX